Reversing
Population Growth
Swiftly and
Painlessly

Reversing Population Growth
Growth
Swiftly and Painlessly

A SIMPLE TWO-CREDIT SYSTEM TO
REGULATE BIRTH RATES AND
IMMIGRATION

4TH EDITION

William Brodovich

Swift Run Press

Swift Run Press
3085 Belvidere St.
Ann Arbor, Michigan 48108
www.reversingpopulationgrowth.com
ISBN: 978-0-9855551-8-4

Cover Illustrations

The *ensō* (circle form) on the front cover is a Japanese symbol of enlightenment. Some *ensō* are closed and others (like this one) have a slight gap.

The back cover depicts a basin for ritual washing—a *tsukubai*. The one shown is located at the Ryōan-ji monastery in Kyoto. Made of stone, it has the shape of an ancient Japanese coin (round with a square hole). A square represents a mouth in Japanese writing. The four characters around the basin's mouth are cleverly designed to be read with the mouth as an integral part of each character. When read clockwise (starting under the bamboo water pipe) the characters say "I only plenty know" (*ware, tada, taru, shiru*). The basin is always full, but never overfull. Its input always equals its output—as in a stable population or a steady-state economy. It knows sufficiency and prevents excess.

The basin was drawn by Ayelén Salinas of San Juan, Argentina.

Μηδέν Ἄγαν ("nothing in excess")

—inscription on the Temple of Apollo, Delphi

Contents

Preface to the First Edition

The insight that population growth could be effectively regulated by a simple two-credit system came to me unbidden in the fall of 2008. I could find no precedent for this idea, so I decided to develop it and get it published. If nothing else, I hope to show that even seemingly intractable problems like overpopulation and poverty could be solved swiftly and painlessly.

Quite a few subjects are covered here, and specialists in each of them will see ways my treatments could be improved. I welcome their comments.

Preface to the Second Edition

The first edition was well received by the few who read it, but I soon realized it needed more work. For one thing, it lacked a detailed description of how the two-credit system would work in a specific country—a model that could be critically examined and challenged. For another, it failed to treat several topics that would have provided more comprehensive coverage of the overpopulation problem. I believe those shortcomings have been remedied in this edition. Readers will now find a detailed description of how the two-credit system would work in the United States and several new chapters covering these topics:

- Why it has become taboo to discuss overpopulation
- Why a voluntary approach to population reduction can never succeed in the long run
- Why reversing population growth would be the most effective way to reverse climate change

- Why the two-credit system would enable us to become the first species to harness natural selection
- Why the world's major religions have ignored overpopulation.

This book benefited greatly from the help of many teachers. Notable among them were John Stuart Mill, Thomas Malthus, Kenneth Boulding, Herman Daly, Michel de Montaigne, Edward Abbey, Aldous Huxley, Henry Thoreau, Charles Darwin, Malcolm Potts, Albert Bartlett, Lucretius, Robinson Jeffers, Thomas Jefferson, Garrett Hardin, Leopold Kohr, Lao Tzu (a collective name), and the team of Donella Meadows, Dennis Meadows, and Jørgen Randers. All of them speak in this book.

If, after reading this book, you feel your time was well spent, please inform others and consider posting a brief review to online booksellers. The sooner people learn about the problem of overpopulation (and how to solve it), the sooner they will demand an end to it.

Preface to the Third Edition

The third edition differs little from the second. I created it to circumvent a publisher who was attempting to claim rights to the book.

Preface to the Fourth Edition

The fourth edition substantially expands and refines the earlier editions. It updates the statistics, adds new material, and expands the bibliography.

Permissions

I thank the following authors, publishers, and owners of copyright for permission to quote passages from the works listed below:

Abbey, Edward. *Confessions of a Barbarian.* Copyright © 1994 by Clarke Abbey. Reprinted by permission of Johnson Books, a Big Earth Publishing Company.

———. *One Life at a Time, Please.* North American print and e-book permission granted by Henry Holt and Company. World print and e-book permission granted by Don Congdon and Associates.

Baker, Dean. *Plunder and Blunder.* Copyright © 2008 by Dean Baker. Reprinted by permission of Berrett-Koehler Publishers, Inc., San Francisco, CA. All rights reserved. www.bkconnection.com

Berry, Wendell. *What Matters?: Economics for a Renewed Commonwealth.* Copyright © 2010 by Wendell Berry. Reprinted by permission of Counterpoint Press.

———. *What Are People For?* Copyright © 1990 by Wendell Berry. Reprinted by permission of Counterpoint Press.

Daly, Herman E. *Beyond Growth: The Economics of Sustainable Development.* Copyright © 1996 by Herman E. Daly. Reprinted as fair use (less than 250 words). Beacon Press books are published under the auspices of the Unitarian Universalist Association of Congregations.

Hardin, Garrett. *Living within Limits.* Copyright © 1993 by Oxford University Press. Reprinted by permission of Oxford University Press, Inc. www.oup.com.

———. *Filters Against Folly*. Copyright © 1985 by Garrett Hardin. Used by permission of Viking Books, an imprint of Penguin Publishing Group, a division of Penguin Random House, LLC. All rights reserved. penguinrandomhouse.com.

Heinberg, Richard. *The End of Growth: Adapting to Our New Economic Reality*. Copyright © 2011 by Richard Heinberg. Reprinted by permission of New Society Publishers.

Huxley, Aldous. *Brave New World Revisited*. Copyright © 1958 by Aldous Huxley. Print book permission from HarperCollins Publishers. E-book permission from Georges Borchardt, Inc. on behalf of the Aldous and Laura Huxley Trust. All rights reserved.

———. *Eyeless in Gaza*. Copyright © 1936 by Aldous Huxley. E-book permission from HarperCollins Publishers. Print permission from Georges Borchardt, Inc., on behalf of Aldous and Laura Huxley Trust. All rights reserved.

Jeffers, Robinson. *The Selected Poetry of Robinson Jeffers*, edited by Tim Hunt. Copyright © by Robinson Jeffers. "November Surf" and the first stanza of "Memoir" reprinted by permission of Stanford University Press. www.sup.org.

Meadows, Donella, with Dennis Meadows and Jørgen Randers. *Limits to Growth: The 30-Year Update*. Copyright © 2004 by Dennis Meadows. Use of less than 250 words is considered fair use by Chelsea Green Publishing.

Montgomery, David R. *Dirt: The Erosion of Civilizations*. Copyright © 2007 by David R. Montgomery. Reprinted by permission of the University of California Press. www.ucpress.edu.

Lucretius. *Lucretius: The Way Things Are, the De Rerum Natura of Titus Lucretius Carus*, translated by Rolfe Humphries. Copyright © 1968 by Indiana University Press. Reprinted by permission of the Indiana University Press.

Introduction

Can you think of any problem in any area of human endeavor on any scale, from microscopic to global, whose long-term solution is in any demonstrable way aided, assisted, or advanced by further increases in population, locally, nationally, or globally?

—Albert Bartlett[1]

The world's population grows by 1.5 million every week (>200,000 every 24 hours) and currently exceeds 8 billion.[2] There are no benefits to this growth, only harm. The proliferation of people ruins environments, cripples economies, and debases the human spirit. Yet politicians never mention overpopulation in their speeches, and journalists almost never write about it. We don't even discuss it privately. The topic has become so taboo that even environmental organizations shun it. Instead, they campaign against the *symptoms* of overpopulation, such as pollution, climate change, species extinctions, and loss of natural areas. They remind me of people on a sinking boat who focus all their energy on bailing instead of first plugging the leak.[3]

Among the world's growing nations, China stands almost alone in having taken decisive measures to curb population growth. In 1980, China established a mandatory one-child policy for about one-third of the Chinese population (mainly city dwellers). This policy spared China some 400 million people and spared the world a commensurate amount of pollution and resource loss.[4] If not for this policy, China today would be unable to feed itself and would lack the capital to develop its economy.

Mandatory birth limits are only one way to curb the birth rate. Another way is to give parents incentives for small families. Singapore did this in the 1970s by offering parents better apartments and better schools in exchange for low fertility. These rewards proved so attractive that the fertility rate fell to only 1.4 children per woman—far below the replacement level of 2.1. This alarmed the Singaporean government, which feared it would lead to a future labor shortage, so it reversed course and began offering educated women incentives to have *more* children.[5]

Birth rates can also be lowered by disincentives, such as a tax on parenthood or the elimination of child tax credits:

> My answer is simple: Place a good stiff tax on Motherhood. Penalize parents. Revise the tax system so as to reward singles and childless couples while requiring the begetters of children (including me) to pay more, not less, in taxes. Economic incentives, if properly designed, should do the trick, thus achieving the end desired without curtailing personal liberties.
>
> —Edward Abbey[6]

Yet another way to lower the birth rate would be to grant people marketable licenses to have children. This idea was proposed half a century ago by the eminent economist and social philosopher Kenneth Boulding:

> I have only one positive suggestion to make, a proposal which now seems so farfetched that I find it creates only amusement when I propose it. I think in all seriousness, however, that a system of marketable licenses to have children is the only one which will combine the minimum of social control necessary to the solution to this problem with a maximum of individual liberty and ethical choice.[7]

Regrettably, Boulding never developed his novel idea into a workable system, confessing that

> ...the sheer unfamiliarity of a scheme of this kind makes it seem absurd at the moment. The fact that it seems absurd, however, is merely a reflection of the total unwillingness of mankind to face up to what is perhaps the most serious long-run problem.[8]

What Boulding called licenses is what today we would call permits or credits (as in pollution permits and carbon credits). Boulding's proposal was a milestone: it was the first time anyone had proposed using tradable credits to solve a social or environmental problem.[9] Few people have ever heard of Boulding's groundbreaking proposal, but two who did were economist Herman Daly and theologian and environmentalist John B. Cobb Jr. They commended it in their book, *For the Common Good.*[10] Boulding's proposal has also been the subject of two favorable professional papers.[11]

This book takes up where Boulding left off. It transforms his idea of using tradable credits to regulate population size into a practical system that would win broad public support. It differs from Boulding's proposal in some important ways. For one thing, it treats credits as a birthright, not something to be earned. And it replaces the awkward 10-, 11-, or 12-credit system suggested by Boulding with a simple two-credit system—a system that would regulate immigration as well as births.

Under the two-credit system, all citizens would have two "population credits" as their birthright. They could use these credits in various ways: they could sell them, donate them, use them to have children (paying two credits for each child), or they could simply let their credits expire when they reach the age of 50. Those wishing to enter the country as immigrants would first have to acquire two credits—either receiving them as a donation or buying them. The government (as the representative of the people) could also purchase credits as needed and retire them or release them in order to regulate population size. Readers will see in Parts 3 and 4 of this book that this two-credit system can gradually reduce any population to an optimal size without imposing a cap on the number of children people can have. Moreover, the plan does not require anyone to violate the moral precepts of any of the world's major faiths regarding birth control and abortion. Another advantage of the plan is that it can be adjusted to overcome population momentum (continued population growth after replacement-level fertility has been achieved due to a large proportion of young people in the population). The two-credit system would provide the following benefits to any society that adopted it:

1. Eliminate all environmental degradation caused by overpopulation, including climate change, deforestation, soil erosion, desertification, wilderness destruction, species extinctions, and contamination of air, water, and soil
2. Eliminate social conflicts generated by overpopulation, preventing many wars and massacres
3. Create a rational immigration policy that would keep the population from growing, and resolve the problem of resident illegal immigrants in a fair and compassionate way
4. Produce a major shift in immigrant demographics away from the poor and poorly educated toward the better off and better educated
5. Greatly reduce the need for social services, policing, courts, and prisons
6. Provide the poor with a means to substantially increase their wealth
7. Largely eliminate teenage births
8. Largely eliminate abortion
9. Greatly reduce child abuse and neglect
10. Greatly reduce the number of foster children
11. Greatly reduce many categories of crime, including rape by strangers
12. Provide all prospective parents with the opportunity to learn the probabilities that their offspring will inherit a

disease or a debilitating defect, thereby giving them the option to reduce suffering and medical expenses

13. Reduce the lethality of disease epidemics due to quicker detection and less crowding

14. Automatically improve the quality of the nation's gene pool without coercion (see "Impact on Population Genetics" in Part 3)

This book is organized as follows: Part 1 lays out the demographic background; Part 2 presents the two-credit plan; Part 3 discusses the demographic changes that would follow implementation of the plan; Part 4 covers various issues that would arise, or should be dealt with, as the population shrinks; Part 5 proposes some political and economic reforms that would complement population reduction; Part 6 proposes health care reforms which would free up money for a program of population control; Part 7 discusses needed ethical reforms; Part 8 examines the hitherto unhelpful role of religion with respect to overpopulation; and Part 9 discusses prospects for implementation of the two-credit plan.

Notes

1. This is Dr. Bartlett's well-known "challenge." Although population growth harms society in the long term, in the short term it helps those who want more consumers for their products, such as manufacturers and homebuilders. In this way, short-term personal gain contributes to long-term social harm.

2. See http://www.geohive.com/earth/population1.aspx.

3. The reasons behind the widespread reluctance to face up to the overpopulation disaster are discussed in a later chapter of this book

("Why the Silence?"). Several other authors have also explored this topic, including Albert Bartlett in his essay "The Massive Movement to Marginalize the Modern Malthusian Message." Bartlett's essay is included in his book *The Essential Exponential! For the Future of Our Planet.*

4. Therese Hesketh, Li Lu, and Zhu Wei Xing, "The Effect of China's One-Child Policy after Twenty-Five Years," *The New England Journal of Medicine,* no. 353 (2005), 1171–76. Can be viewed online. Interestingly, even before the implementation of the one-child policy, China's total fertility rate fell by half between 1970 and 1979. This was due to the voluntary "late, long, few" policy (marry late, space children far apart, have few of them). In 2015, China replaced the one-child policy with a two-child policy. The two-child policy caused the birth rate to rise 7.9 percent by the end of 2016. (*Time Magazine*, February 6, 2017, p. 12). In May 2021 China adopted a three-child policy.

5. John R. Weeks, *Population: An Introduction to Concepts and Issues*, 10th ed. (Thomson Wadsworth, 2005, 2008), 513–514.

6. Edward Abbey, "A San Francisco Journal" in *One Life at a Time, Please* (New York: Henry Holt & Co., 1987), 66–67.

7. Kenneth Boulding, *The Meaning of the Twentieth Century: The Great Transition* (New York: Harper & Row, 1964), 135.

8. Ibid., 136.

9. Herman Daly, *Beyond Growth: The Economics of Sustainable Development* (Boston: Beacon Press, 1996), 56: "It [a system of tradeable permits] can even be applied to population control as in the tradeable birth quotas suggested by Kenneth Boulding (1964). In fact, to my knowledge, Boulding's was the first clear exposition of the logic of the scheme, although applied to the least likely area of acceptance politically."

10. Herman Daly and John B. Cobb Jr., *For the Common Good: Redirecting the Economy toward Community, the Environment, and a Sustainable Future* (Boston: Beacon Press, 1994), 244–46.

11. David M. Heer, "Marketable Licenses for Babies: Boulding's Proposal Revisited," *Social Biology* 22, no. 1 (Spring 1975) and David de la Croix and Axel Gosseries, "Population Policy through Tradable Procreation Entitlements," *Society for the Study of Economic Inequality* (1975). Available online at http://www.ecineq.org/milano/WP/ECINEQ2007-62.pdf.

Part I:

The Problem of Overpopulation

Defining Overpopulation

To the size of states there is a limit, as there is to other things, plants, animals, implements; for none of these retain their natural power when they are too large or small, but they either wholly lose their nature or are spoiled.

—Aristotle[1]

Overpopulation means *too many people*. *Too many* means the population has exceeded its optimal size. The optimal size is the size that would optimize our potential for happiness without unduly infringing on the well-being of other creatures. It is a size that represents foresight, self-restraint, generosity, and empathy. It is biocentric, not anthropocentric.[2]

A thing is right when it tends to preserve the integrity, stability, and beauty of the biotic community. It is wrong when it tends otherwise.

—Aldo Leopold[3]

Most ecologists use another definition of overpopulation. They define overpopulation as occurring when the population exceeds the size that can be sustained indefinitely. This is the concept of carrying capacity. It differs from the first definition in this respect: carrying capacity is the *largest* sustainable population size, while the optimal size will always be much smaller.

Just as we can temporarily live beyond our means by going into debt, nations can temporarily exceed their carrying capacity by consuming the carrying capacity of future generations. All nations do that. They consume resources faster than they are replenished and produce pollution faster than nature can process it. This not only deprives future generations of needed resources but destroys the carrying capacity of other species.

Thomas Malthus, in his influential book *An Essay on the Principle of Population* (1798), was the first to examine in depth the problem of overpopulation. (He called it "redundant population.") Malthus defined overpopulation as exceeding "the means of subsistence" or "the limits of the food." In other words, he defined overpopulation as exceeding the limits of sustainability. He made no distinction between short-term and long-term sustainability, because environmental degradation was poorly understood in his time.

Although Malthus believed that people living in poverty should not reproduce, he favored population growth in general. He believed population growth was the natural course of civilization. He also believed that once a population grows large, the competition for resources will spur technological innovations that will allow even more population growth. But he cautioned that these technological advances can never keep pace with a rapidly expanding population.

The reason Malthus favored population growth was his concern that if men did not have large families to support, they would become lazy, and their nation would end up too thinly populated and vulnerable to conquest. He felt that small families would remove a "motive sufficiently strong to overcome the acknowledged indolence of man, and make him proceed in the cultivation of soil. The population of any large territory, however fertile, would be as likely to stop at five hundred, or five thousand, as at five millions, or fifty millions."[4]

Charles Darwin, whose reading of Malthus was key to the development of his theory of evolution, also believed population growth was desirable. Darwin's rationale was that population growth fosters intense competition which serves to make *Homo sapiens* fitter over time. In *The Descent of Man,* he said:

> Man, like every other animal, has no doubt advanced to his present high condition through a struggle for existence consequent on his rapid multiplication; and if he is to advance still higher he must remain subject to a severe struggle…Hence our natural rate of increase though leading to many and obvious evils, must not be greatly diminished by any means.[5]

John Stuart Mill, in contrast to both Malthus and Darwin, believed that a stable population was desirable. Mill was also careful to distinguish between the maximum sustainable population size (carrying capacity) and the optimal size:

> There is room in the world, no doubt, and even in the old countries, for a great increase in population, supposing the arts of life to go on improving, and capital to increase. But even if innocuous, I confess I see very little reason for desiring it. The density of

population necessary to enable mankind to obtain, in the greatest degree, all the advantages both of co-operation and of social intercourse, has, in all the most populous countries, been attained. A population may be too crowded, though all be amply supplied with food and raiment. It is not good for man to be kept perforce at all times in the presence of his species. A world from which solitude is extirpated, is a very poor ideal. Solitude, in the sense of being often alone, is essential to any depth of meditation or of character; and solitude in the presence of natural beauty and grandeur, is the cradle of thoughts and aspirations which are not only good for the individual, but which society could ill do without. Nor is there much satisfaction in contemplating the world with nothing left to the spontaneous activity of nature; with every rood of land brought into cultivation, which is capable of growing food for human beings; every flowery waste or natural pasture plowed up, all quadrupeds or birds which are not domesticated for man's use exterminated as his rivals for food, every hedgerow or superfluous tree rooted out, and scarcely a place left where a wild shrub or flower could grow without being eradicated as a weed in the name of improved agriculture. If the earth must lose that great portion of its pleasantness which it owes to things that the unlimited increase of wealth and population would extirpate from it, for the mere purpose of enabling it to support a larger but not a better or happier population, I sincerely hope, for the sake of posterity, that they will be content to be stationary, long before necessity compels them to it.[6]

The writer Edward Abbey also pondered the question of optimal population size and the related question of optimal economic development:

What the conscience of our race—environmentalism—is trying to tell us is that we must offer to all forms of life and to the planet itself the same generosity and tolerance we require from our fellow humans. Not out of charity alone—though that is reason enough—but for the sake of our own survival as free men and women. Certainly the exact limits of what we can take and what we must give are hard to determine; few things can be more difficult than attempting to measure our needs, to find that optimum point of human population, human development, human industry beyond which the returns begin to diminish. Very difficult; but the chief difference between humankind and the other animals is the ability to observe, think, reason, experiment, to communicate with one another through language; the mind is our proudest distinction; the finest achievement of our human evolution. I think we may safely assume that we are meant to use it.[7]

Abbey even suggested a formula for determining an optimal population size:

> …how much Nature is enough? Enough to go around, I'd say, or about one square mile per human—with a little surplus left over.[8] By a "little surplus" I mean wild areas where through general agreement none of us enters at all. An absolute wilderness, we might call it, justified by our recognition of the rights of other living things to a place of their own, a role of their own, an evolution of their own not influenced by human pressures. A recognition, even, of the right of nonliving things—boulders, for example, or an entire mountain—to be left in peace, alone, for a few centuries now and then. A foolish, utopian idea, no doubt; I advance it merely as a suggestion of what is possible were the human consciousness, and the human conscience, ever to reach so generous a level. It is not enough to understand the natural world; the point is to preserve it. Let Being be.[9]

In the end, it is not we, but our descendants, who will determine the optimal population size for our nation and our local communities. Our responsibility now is to halt further growth and begin reducing our population to a size large enough for security and a rich cultural life but small enough to allow coexistence with all the other creatures that share our planet.[10]

Notes

1. Aristotle, *Politics*, trans. Benjamin Jowett, part IV. Available online at http://classics.mit.edu/Aristotle/politics.7.seven.html.

2. A practical way to monitor our impacts upon other species would be to do so on a watershed basis. For each watershed (or subwatershed), a panel of biologists (selected by other biologists) would monitor the welfare of other species and determine when significant harm was starting to occur. At that point, the human population within that watershed would have to rein in whatever practices were causing the harm or reduce its numbers.

3. *A Sand County Almanac and Sketches Here and There* (Oxford University Press, 1949), 224–25.

4. Donald Winch, ed., *An Essay on the Principle of Population*, (Cambridge University Press, 1992), 214.

5. *The Descent of Man*, chapter 21. Malthus believed chastity was the proper way to reduce birth rates. He opposed artificial contraception and would therefore have opposed the two-credit system. Darwin, on the other hand, would likely have endorsed the system. He would have recognized that it would protect biodiversity and simultaneously improve the human genome. He would also have understood that the two-credit system does not eliminate competition, but merely civilizes it, thereby preventing the "many and obvious evils" he deplored.

6. John Stuart Mill, *Principles of Political Economy* (New York: Prometheus Books, 2004), 691–92.

7. Edward Abbey, *Abbey's Road* (New York: Penguin Books, 1979), 135.

8. Jared Diamond in his book *The Third Chimpanzee* (p. 344) reckoned that populations of modern hunter-gatherers even in the best habitats number only about one per square mile. Thomas Jefferson came to the same conclusion: he estimated that the original

population density of the Powhatan Indians of Virginia was one per square mile. (See Jefferson's *Notes on the State of Virginia*, Query XI.)

9. Edward Abbey, *Down the River* (New York: A Plume Book, 1982), 119.

10. Darwin and Malthus (both of whom favored population growth) had 10 and 3 children respectively. John Stuart Mill, who favored a stable population, had 1 (adopted) daughter. Edward Abbey and Aldo Leopold each had 5 children, while Garrett Hardin had 4. Abbey liked to point out (with a grin) that he averaged only 1 child per wife, urging others to follow his example.

The Arithmetic of Growth

To understand population growth, we need to understand the math of growth. Happily, this is not difficult. The first step is to distinguish between *linear* growth and *exponential* growth. Linear growth occurs when something grows steadily, such as a strand of hair or money deposited weekly in a jar. When we graph linear growth, it is always a straight diagonal line (hence its name). Exponential growth, in contrast, is growth that accelerates over time. It does so because it is based on a constant percentage of the total. When we graph exponential growth, the line initially rises slowly, then gains speed, and finally soars nearly vertically. Malthus called exponential growth the *principle of population.*

Some examples of exponential growth are the multiplication of neutrons in nuclear chain reactions, the growth of money in interest-bearing accounts, the reproduction of organisms (both sexually and asexually), and the spread of viral videos. Even the cells that comprise our bodies multiply exponentially until we mature.

To appreciate the impact of exponential growth, we need to know how to compute doubling time. To do this, we simply multiply the natural log of 2 (which is 0.693) by 100, and then divide the product (69.3) by the annual percent growth rate. (The reason we multiply the log of 2 by 100 is that we are dealing with a *percent* growth rate.) If we round 69.3 to 70, we can do many doubling-time calculations in our head. For example, if something is growing at 2 percent annually, it will double in 35 years (70 divided by 2). If it is growing at 5 percent annually, it will double in 14 years (70 divided by 5). The US population in 2021 was growing at 0.7 percent annually. If that rate continues, the population will double in 100 years. If we want greater accuracy in our calculations, we can use an online compound interest calculator.

To illustrate the remarkable power of exponential growth, let's observe how tall a child would be by the age of 21 if the child retained the same 52 percent growth rate that he or she had during the first year of life.[1] To calculate the child's height at different ages, we will use an online compound interest calculator, and assume an initial height ("principal") of half a meter. These online calculators allow us to choose how many times per year we want the compounding to occur. In our example, we'll choose 12 times (monthly). The following table shows the results:

Child's Age (in years)	Height (in meters)
1	0.5
2	1.38
5	6.37
10	81.23
15	1,035.44
21	21,957.70

By the time the child is 10 he would have the height of a 23-story building, and by the age of 21 his height would be 2.5 times the height of Mt. Everest (whose summit is 8,848 meters above sea level). If he were to keep growing at this rate for an average American life span of 79 years, his final height would be 145.9 quintillion meters (equivalent to 15,400 light-years). To put that in perspective, the nearest star (other than the sun) is 4.4 light-years away, and our Milky Way galaxy is 100,000 light-years across.

Fortunately, nature slows the human growth rate after the first year and eventually stops it altogether. Alas, nature does not do the same for population growth. But nature compensated us by giving us a brain that can recognize the dangers of overpopulation and take measures to prevent or reverse it.

In the real world, populations don't grow at steady rates but at rates that change from year to year and decade to decade. We can think of such growth as variable-rate exponential growth. It is synonymous with variable-rate compound interest.[2] If we think of a population as principal and think of offspring as interest, the parallel between the growth of money and the growth of populations becomes clear.

Humans have always found it difficult to grasp the magnitude and implications of exponential growth. This has led to unrealistic thinking about our planet's capacity to support human life. Listen, for example, to William Godwin, a prominent 18th-century social reformer:

> Three-fourths of the habitable globe is now uncultivated. The parts already cultivated are capable of immeasurable improvement. Myriads of centuries of still increasing population may pass away, and the earth be still found sufficient for the subsistence of its inhabitants.[3]

Godwin's cornucopian fantasy was echoed in our own time by Julian Simon, an economist and professor of business administration at the University of Maryland:

> We now have in our hands, in our libraries, really the technology to feed, clothe, and supply energy to an ever-growing population for the next 7 billion years...We [are] able to go on increasing forever.[4]

Simon's rosy predictions display not only his stunning ignorance of exponential growth, but as Ernest Partridge pointed out, "Missing from Simon's cheerful prognoses is any acknowledgment or apparent comprehension of such fundamental ecological principles as nutrient cycling, feedback mechanisms, and limiting factors, or even that very foundation of physical science: thermodynamics and entropy. His perspective is confined to his own field of market economics."[5] Moreover, Simon was clearly unaware that in less than 4 billion years, life on earth will be impossible because the sun will be far too hot.

Sadly, Simon's ignorance is shared by most of the world's leaders. The members of the US Congress, for example, did not foresee that inserting a "family reunification" provision into the 1965 immigration bill would trigger an exponential increase in immigration by allowing immigrants to sponsor their relatives for immigration and allowing those relatives to sponsor even more relatives.[6] Instead of *e pluribus unum* (from many, one), Congress gave us *ex uno plures* (from one, many).[7]

Notes

1. Baby boys in America are about 50 cm long when born and about 76 cm long after one year. They therefore have a first-year growth rate of 52 percent (26 cm/50 cm × 100 = 52).

2. Math textbooks simplify their descriptions of exponential growth by using a fixed growth rate, which has misled some to think that exponential growth requires a fixed growth rate. For an informative discussion of variable-rate exponential growth, see "Albert Bartlett: An Exponentialist View" by David A. Coutts (2006) at http://members.optusnet.com.au/exponentialist/Bartlett.htm.

3. Godwin as quoted by Thomas Robert Malthus in *An Essay on the Principle of Population* (1798), chapter X. This is page 76 of the Oxford World Classics edition and page 58 of the Cambridge University Press edition.

4. Norman Myers and Julian Simon, *Scarcity or Abundance?: A Debate on the Environment* (New York: Norton, 1994), 65.

5. Earnest Partridge, "Perilous Optimism" (February 2001) http://gadfly.igc.org/papers/cornuc.htm.

6. According to the National Immigration Forum (https://immigrationforum.org), the average US immigrant who has permanent residency or citizenship petitions to have 3.5 relatives come to the US. Family reunification accounts for 65 percent of legal US immigration.

7. To learn more about the background of the family reunification provision in the 1965 Immigration and Nationality Act (and the failure of Congress to correct its mistake), see chapter 11, "The Accidental Betrayal," in Roy Beck's fine new book *Back of the Hiring Line*.

Recent Unprecedented Population Growth and its Implications

The tremendous increase in the global human population in the last 180 years (two human lifespans) is shown graphically below. From approximately 1 billion people in 1830, the world's population has swelled to 8 billion in 2022. It took 100 years to get from 1 billion to 2 billion; just 30 years to reach 3 billion; 14 years to reach 4 billion; 12 years to reach 5 billion; 14 years to reach 6 billion; 11 years to reach 7 billion (in 2011); and another 11 years to reach 8 billion (in 2022). The increase in the world's population in the decade 2000–2010 exceeded the number of people present on the Earth in 1600. At its current growth rate of 0.84 percent, the world's population will double in 83 years, swelling to more than 16 billion by 2106. But the world is unlikely to ever suffer that many people, because the resources simply don't exist to support them.

Growth of World Population (1830-2020)

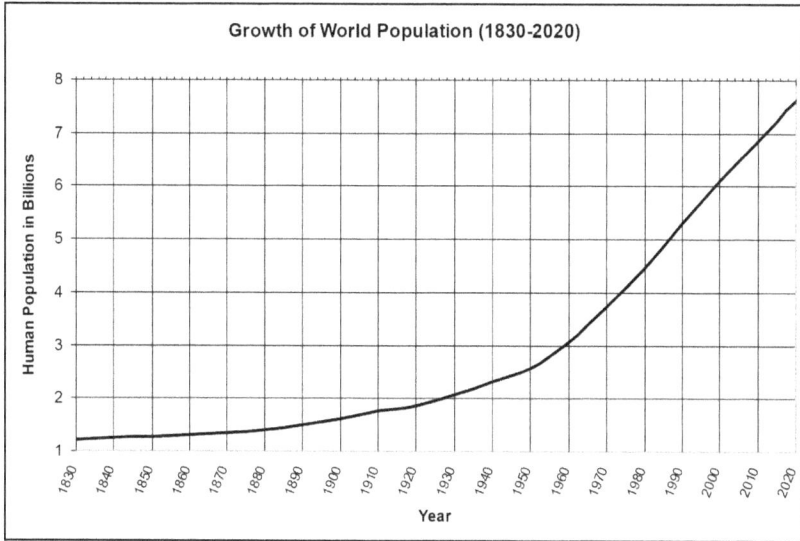

Population growth in the United States has paralleled that of the rest of the world, rising from about 13 million in 1830 to 335 million in 2022—a 26-fold increase in two human life spans. Only China and India have larger populations than the United States. How long it will take the US population to double will largely depend on the rate of immigration, because immigration is the prime driver of US population growth.[1]

US Population Growth (1830-2020)

Population (millions) vs. Year

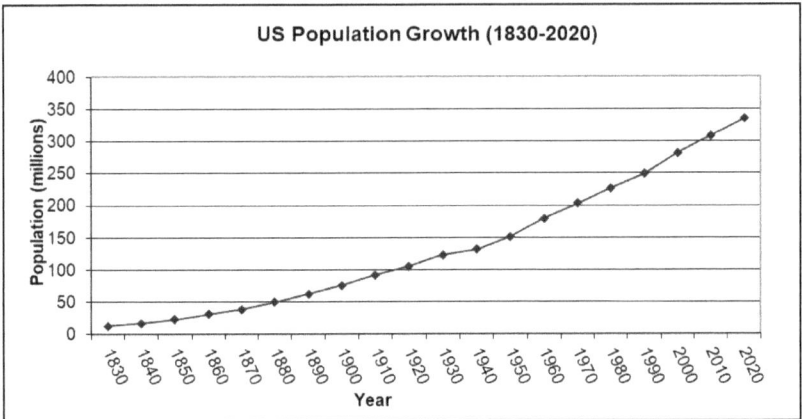

The extraordinary increase in world population was brought about by technological advances in sanitation, immunization, synthetic fertilizer, plant breeding, antibiotics, water purification, and mosquito control. As mortality rates fell, birth rates, unfortunately, remained high, causing rapid population growth. These high fertility rates are now declining in many places.

Some believe mankind can adapt to high rates of population growth by developing new resources, using resources more efficiently, distributing resources more equitably, and finding more effective ways to cope with pollution. They point to improvements in medicine, sanitation, agriculture, and global charity that have so far averted a major population collapse. They also note that the rate of world population growth has slowed from a high of 2 percent in the 1960s to 0.84 percent today.[2] In many countries, birth rates have even fallen below the replacement level.

Others are more pessimistic about population growth. They note that even in countries where birth rates are below replacement level, populations are often still growing due to population momentum and/or immigration. They point to alarming signs of ecological stress and note that every species that overpopulates eventually has its growth halted by famine, epidemics, war, and impaired fertility. They reject claims that population growth boosts economies by pointing to the impoverished economies of overpopulated countries like Afghanistan, Haiti, and Rwanda.

The principal reason nations with rapidly growing populations nearly always have poor economies is that parents who have many children must devote all their income to meeting basic needs and paying down debt. They have no savings (capital) left over to invest. Without capital, an economy cannot grow fast enough to generate the new jobs needed by the growing population.[4] Without jobs, young people are forced to emigrate or stay behind to fight over the scraps. To appreciate the scale of the problem, consider Nigeria: The number of young Nigerian men entering the job market is currently seven times greater than the number who are vacating jobs through death or retirement.[5] That means the Nigerian economy must generate six new jobs for each worker who retires or dies—an impossible task. And even when people succeed in getting a job, they are paid very little, because so many people are desperate for work.

> A market overstocked with labour, and an ample remuneration to each labourer, are objects perfectly incompatible with each other. In the annals of the world they never existed together; and to couple them, even in imagination, betrays a gross ignorance of the simplest principles of political economy.
>
> —Thomas Malthus[6]

Accompanying the rapid increase in the world's population has been a huge increase in the proportion of people living in cities. More than half of the world's people now live in urban areas.[7] Much of this urban growth is due to migration from overpopulated rural areas and is perpetuated by high birth rates and relatively low death rates within cities.[8] At least one-third of the world's city dwellers live in squalid slums, where they are deprived of all contact with nature.[9] Not surprisingly, they suffer high rates of mental illness. For example, schizophrenia is 30 percent more common in cities than in rural communities.[10]

As populations expand, so too does the amount of regulation needed to manage the ever-increasing social and environmental complexity. In an optimally populated world, we would need relatively few regulations beyond those necessary to regulate birth rates and immigration. For this reason, those who decry big government should be at the forefront of those calling for population reduction.

Another downside of population growth is that it reduces people's political power. If we compare the representation of enfranchised citizens in the United Sates in 1790 (807,095) to the number of enfranchised US citizens today (about 196,899,193), we find that each of the 66 congressmen during George Washington's presidency represented an average of 12,229 voters whereas each of today's 435 congressmen represents an average of 452,642 voters. We can visualize this loss of personal political power by comparing it to body height: If enfranchised Americans stood 6 feet tall during George Washington's administration, they now stand 1.9 inches tall and are shrinking rapidly.[11] The House of Representatives would have to be expanded from its present 435 members to 16,095 members to recapture the full political influence that enfranchised citizens enjoyed when Washington was president. But the situation is even worse than that, because the wealthy have been able to overcome the handicap of being only 1.9 inches tall by funding political campaigns. This has transformed them into giants while further diminishing the rest of us. By allowing our population to grow too large (and failing to control campaign financing), we have made ourselves politically impotent.

> The world's oldest and most stable democracies are all small and local. There is Iceland with its three hundred thousand inhabitants. There is the Isle of Man with a population of eighty-five thousand, whose parliament, the Tynwald, is likewise of Norse origin, and likewise over a thousand years old. There is Switzerland, which has eight million inhabitants but is a confederation of twenty-six cantons, the largest of which, Bern, has a population of under a million, so that no one canton can dominate the rest. Beyond a certain size it becomes impossible to get anything done without acting through representatives, and that is where the rot sets in. To work as a direct democracy, each citizen has to have a sizeable share of sovereignty.
>
> —Dmitry Orlov[12]

Notes

1. See the "US POPClock Projection" at the US Census Bureau website: https://www.census.gov/content/dam/Census/library/publications/2020/demo/p25-1146.pdf.

2. https://www.worldometers.info/world-population/

3. Perpetual growth is only possible on a flat earth, since only a flat earth can extend infinitely. A sphere, in contrast, always has a boundary and is therefore finite. See Albert Bartlett, "The Exponential Function XI: The New Flat Earth Society," in *The Physics Teacher* 34, no. 6 (September 1996), 342–343. Bartlett's article is included in *The Exponential Function! For the Future of Our Planet* (see bibliography).

4. John R. Weeks, *Population: An Introduction to Concepts and Issues*, 6th ed. (Wadsworth Publishing Company, 1996), 441.

5. Ibid., 449.

6. *An Essay on the Principle of Population*, edited by Donald Winch. Cambridge University Press, 1992, 231; *UN State of the World's Cities Report 2008/2009.*

7. Different countries use different definitions of "urban," so the percentage of the world's population that lives in urban areas can only be estimated.

8. *UN State of the World's Cities Report*, 39.

9. John R. Weeks, *Population: An Introduction to Concepts and Issues*, 10th ed. (Thomson Wadsworth, 2008), 379.

10. Florian Lederbogen, Peter Kirsch, Leila Haddad, Fabian Streit, Heike Tost, Philipp Schuch, Stefan Wust, Jens C. Pruessner, Marcella Rietschel, Michael Deuschle, and Andreas Meyer-Lindenberg, "City Living and Urban Upbringing Affect Neural Social Stress Processing in Humans," *Nature* 474: 498–501. Published online June 22, 2011. Another factor besides stress in the etiology of urban schizophrenia may be the cat-borne protozoan *Toxoplasma gondii*. See "How Your Cat Is Making You Crazy" by Kathleen McAuliffe in the March 2012 issue of *The Atlantic*. This protozoan normally passes between cats and rodents, both of which are plentiful in urban slums.

11. Here is the arithmetic: 72 inches/1.9 inches equals 452,642 voters/12,229 voters.

12. Dmitry Orlov, *The Five Stages of Collapse* (Gabriola Island, B New Society Publishers, 2013), 67.

Why the Silence?

Responsible couples always discuss how many children they should have and take the necessary precautions to ensure they don't exceed that number. But as a people, we never discuss how large our nation should be. At most we may discuss whether illegal immigrants should be legalized or how many refugees to admit. We are so unaccustomed to thinking about the impacts of population growth that we don't even discuss how large our local communities should be. Instead, we let them metastasize until we no longer enjoy living in them.[1]

The main reason people are reluctant to talk about overpopulation is their fear of controversy. Any discussion of mankind's biggest problem inevitably touches on "hot-button" issues such as abortion, immigration, contraception, women's rights, parental rights, reproductive rights, and religion. Views on these subjects tend to be strongly held and polarized. Some examples:

A: "*Undocumented workers* should be legalized."
B: "*Illegal aliens* should be deported."

A: "Abortion should be legal and readily available."
B: "Abortion is murder."

A: "Contraception empowers women, giving them control over their lives."

B: "Contraception is immoral."

A: "Parents should determine family size."

B: "God should determine family size."

C: "Society should determine family size."

A: "We should admit more immigrants out of compassion."

B: "We should exclude more immigrants because we need to conserve resources, reduce pollution, and protect our jobs."

A: "We should admit more immigrants because we need more consumers to grow the economy."

B: "We need to halt both economic growth and population growth in order to slow the depletion of nonrenewable resources and reduce pollution."

A: "We can accommodate ever more people thanks to never-ending technological innovation."

B: "We are already living unsustainably. Technology has never kept up and never will."

A: "We can accommodate ever more people thanks to God's boundless providence. With God all things are possible."

B: "God helps those who help themselves."

C: "There is no God."

A: "We should admit more immigrants because our ancestors were immigrants."

B: "Our ancestors immigrated because their parents had too many children. Overpopulation is the driver of immigration."

Another reason people are reluctant to speak up about overpopulation is their fear of being taken for misanthropes. They worry that others will wrongly conclude that they scorn babies, despise motherhood, lack compassion for desperate refugees, and resent the parents of large families.[2] One of the finest features of the two-credit system is that it would allow us to celebrate *all* births, welcome *all* legal immigrants, and preclude *all* resentment of large families, because the addition of these children and immigrants could never cause the population to grow. In fact, the population would gradually shrink to an optimal size even as those babies and immigrants were added. Children and refugees have no better friends than the opponents of overpopulation, for it is they who understand that overpopulation creates more refugees and more suffering for children.

So strong is the taboo against discussing overpopulation that even environmental organizations observe it. Decades ago, these organizations actively campaigned against population growth, but no more. Now all they do is campaign against the *symptoms* of overpopulation, such as climate change, loss of natural areas, collapse of biodiversity, and pollution. The main reason leaders of environmental organizations avoid confronting overpopulation is their fear of offending major donors. The Sierra Club is a prime example. In 1994, the Sierra Club's largest donor, David Gelbaum, warned the Club's executive director, Carl Pope, that if the Club took a stand against mass immigration, he would halt his generous donations.[3] The Club's leadership quickly responded by changing its immigration policy. The old policy said that immigration should not contribute to population growth, but the new policy was neutral on immigration. In 2013, the Club's Board of Directors went even further by endorsing amnesty for America's 11 million illegal immigrants. Amnesty would allow these illegal immigrants to become American citizens, and that would allow them to sponsor millions of their relatives for immigration to the US. The Sierra Club's pro-mass-immigration policy obliges it to absurdly pretend that population growth is compatible with the Club's professed goals of reducing pollution, reducing resource consumption, reducing the loss of natural areas, and halting species extinction.[4]

The US Green Party, like the big environmental organizations, has little to say about population. The Green Party's population platform tamely recommends the following:

- Universal reproductive education
- Ready availability of inexpensive contraception (including the morning-after pill)
- More research to develop better contraceptives
- Keep abortion legal

Crucially, the Green Party's platform does not call for halting—let alone reversing—US population growth. Instead, like the Sierra Club, the Green Party calls for legalizing America's 11 million illegal aliens—something that would lead to a great increase in the nation's population in the absence of an effective system of population control.

Even the prestigious American Association for the Advancement of Science (AAAS) rejects discussion of overpopulation. This organization publishes the highly respected journal *Science*. AAAS does not allow mankind's most serious problem to be discussed in the pages of *Science* or at any of its conferences. Over the past few years, four population-focused NGOs have requested exhibitor booth space at AAAS conferences. In every case, the AAAS has refused. In preparation for the February 2016 conference, Scientists and Environmentalists for Population Stabilization (SEPS) requested booth space and accompanied their request with supportive letters from 40 present and former heads of American scientific societies.[5] Yet once again, the AAAS refused. To put that in perspective, 19 other scientific societies had previously welcomed the SEPS booth at their meetings during the previous four years.

In the United States (where 88 percent of current population growth is due to immigrants and their children),[6] the mass media strongly support continued large-scale immigration. They do so because the media are owned by the business elite, who profit from the lower wages and greater consumption brought about by mass immigration. Their pro-immigration message insinuates that those who oppose mass immigration are anti-immigrant, xenophobic, and racist. That serves to further suppress criticism of mass immigration, because nobody wants to be tarred with such epithets.[7] The truth is that very few opponents of mass immigration are opposed to all immigration. They simply want reasonable limits on immigration, so that it does not contribute to further population growth or otherwise harm society. Those who have already immigrated have no better allies than those who oppose further mass immigration. Adding tens of millions of additional workers would only result in lower wages, increased unemployment, more pollution, more traffic congestion (US commuting times have increased 20 percent since 1980),[8] higher housing costs, more crime, more rapid global warming, and less social cohesion—all of which reduce the quality of life for both immigrants and natives.

The writer Edward Abbey experienced firsthand the adamant refusal of the US mainstream media to allow any serious discussion of the role of immigration in overpopulation. Abbey was asked to write an essay on immigration for the op-ed page of the *New York Times*. He and the *Times* editor agreed on how long it should be. Abbey then wrote the essay and submitted it. After a long delay, the *Times* asked him to cut its length in half. Abbey did so. After another long delay, Abbey contacted the editor to find out what the problem was. He was informed that the essay would not be published for "lack of space." Abbey then requested that the *Times* return his manuscript and pay him the customary kill fee (about $400). The *Times* never responded. Abbey then submitted his essay to *Harper's*, *The Atlantic*, *The New Republic*, *Rolling Stone*, *Newsweek*'s "My Turn," and *Mother Jones*. None would print it. Abbey's essay—which is excellent—was eventually published in his own book, *One Life at a Time, Please*, under the title "Immigration and Liberal Taboos."[9]

In recent years, the mass media have begun feeding us stories aimed at convincing us that the real problem is not overpopulation, but *under*population. The authors of these tales tell us that our current low birth rates will result in a shortage of young workers, making it difficult to fund social pension programs, care for the elderly, or even harvest crops. (They don't seem to realize that young workers will also grow old and require support.) The authors also warn that if the number of consumers shrinks, so will the economy. This is supposed to frighten us, even though a shrinking economy would be a good thing.

A fine illustration of the way the mass media distort population coverage is an article published in *Time* magazine by Hannah Beech (December 2, 2013), entitled "Why China Needs More Children: After Decades of the One-Child Policy Beijing Wants Its People to Have More Kids. It May Be Too Late for That." This story advises readers that China's low fertility rate (which is now well below the replacement level) will lead to a shortage of young workers who are needed for economic growth (assumed to be desirable) and to provide support for the elderly. The author neglects to point out that the fewer children people have, the more they can save for their own health care and retirement. Alternatively, they can afford to pay higher taxes to have the government provide these services. Nor does she mention that the one-child policy, by sparing China more than 400 million people, enabled China to feed itself and have enough capital left over to develop its economy. The author does concede that a shrinking Chinese labor supply will cause wages to rise (and in fact they are already rising). But if a smaller population leads to higher wages, why would the Chinese want more people? Clearly, this author is simply expressing the prevailing ideology of the business elite, who want lower wages and ever more consumers for their unnecessary products and services. For a more recent article echoing the same foolish message, see "China's Twilight Years: As Immigrants Replenish America, China's Population Is Aging and Shrinking" by Howard French in *The Atlantic*, June 2016, 15–17.

A related failure of the mainstream media is that their news stories pertaining to war, migration, famine, and terrorism are never presented in a demographic context. Such stories should always be prefaced with the demographic background of the affected country. That would enable the public to understand that the root cause of the world's problems is reckless breeding.

Ecologist Garrett Hardin pointed out that the population taboo, like all taboos, has a dual nature: There is the taboo itself, and then there is the taboo against acknowledging the taboo.

> The double nature of taboo has not been generally recognized, but a little thought shows that this bivalence is necessary for the stability of a taboo.[10]

The good news is that ignoring overpopulation won't be possible much longer. The problems generated by too many people are becoming too severe to ignore. Like alcoholics, the world's nations (and their environmental and scientific organizations) are going to have to decide whether to continue down the path of destruction or admit the problem and take the necessary measures to correct it. Although it is not too late to save the good things that are left in this world, both wisdom and political will are rare today. If people fail to act, nature will step in with painful consequences. In fact, millions of people are already feeling the pain as one nation after another descends into chaos and violence, triggering mass migrations.

Notes

1. Although most people can readily comprehend the need to limit the size of their own family, many fail to see the equivalent need to regulate the size of their local community or nation. This inconsistency has its exact counterpart in economics: All economists recognize that "in microeconomics every enterprise has an optimal scale beyond which it should not grow. But when we aggregate all

microeconomic units into the macroeconomy, the notion of an optimal scale, beyond which further growth becomes antieconomic, disappears completely!" (Herman Daly, *Beyond Growth: The Economics of Sustainable Development*, p. 27).

2. If two-child families had been the norm in the past, we might never have known the contributions of such luminaries as Charles Darwin (5th child), Alfred Russell Wallace (7th child), Thomas Malthus (7th child), Nicolaus Copernicus (4th child), Thomas Edison (7th child), Thomas Jefferson (3rd child), and William Shakespeare (3rd child). By the same token, two-child families would have spared the world many bad people, such as Adolf Hitler (4th child), Napoleon Bonaparte (4th child), Mao Zedong (3rd child), and Vladimir Putin (3rd child). In any case, had the likes of Darwin, Jefferson, and Shakespeare never existed, others would have taken their place and made their own notable contributions.

3. See Kenneth R. Weiss, "The Man Behind the Land," *Los Angeles Times*, October 27, 2004. Gelbaum said: "I did tell Carl Pope in 1994 or 1995 that if they ever came out anti-immigration, they would never get a dollar from me."

4. Here is the Sierra Club's official position on mass immigration to the United States:

    > Restricting immigration to the United States won't solve the environmental problems that force people to move in the first place, and the increasing numbers of illegal immigrants indicate that restrictions are more thumb-in-the-dike than viable policy. The Sierra Club's international efforts go to the headwaters, promoting environmentally sustainable livelihoods that keep forests and families healthy, while making polluting multinational corporations accountable and trade agreements fair.
    >
    > —Stephen Mills, director of the Sierra
    > Club's international programs.

Two things are striking about the Sierra Club's immigration position: First, the author tells us that foreigners are driven to migrate by "environmental problems." What he neglects to point out is that these environmental problems are the direct result of overpopulation. Second, the author dismisses "thumbs in the dike" (border security, immigration quotas) as "not a viable policy." So, what is the Sierra Club's viable policy? Apparently, it consists of lowering the ocean so that dikes aren't needed. And how does the Sierra Club propose to lower the ocean? Here's how: by endorsing "sustainable livelihoods" over there! Laugh or cry?

My criticism of the leadership of the major environmental organizations does not extend to their active members. As Edward Abbey observed, these are the people "who actually stand before the bulldozers, spike the trees, lobby the politicos, write the tedious letters, lick stamps, staple leaflets, organize committees, attend meetings, hire lawyers and sometimes go to jail, do what they do with no fame, no public credit, certainly little or no pay (except Sierra Club bureaucrats etc.), and no reward but the sense of having opposed the rich and powerful in the name of something more ancient and beautiful than human greed and human increase." (Edward Abbey in a letter to Barry Lopez dated June 14, 1987, in *Postcards from Ed*, Milkweed Editions, 2006, p. 214.)

5. See "AAAS Wields the Censor's Hammer on Population Issues" by Stuart Hurlburt, CAPS blog, February 11, 2016, www.capsweb. org/blog/aaas-wields-censors-hammer-us-population-issues. Also viewable at www.progressivesforimmigrationreform.org/ aaas-wields-the-censors-hammer-on-u-s-population-issues/.

6. See http://www.progressivesforimmigrationreform.org/pfirs-environmental-impact-statement-part.

7. Although advocates for an end to mass immigration are sometimes accused of being anti-immigrant, xenophobic, and racist, advocates for small families are rarely accused of being antichild, child-phobic, and child-hating. This inconsistency is likely due to the fact

that advocates for mass immigration are better funded and more powerful than advocates for more babies.

8. Christopher Ingraham, "The Astonishing Human Potential Wasted on Commutes," *Washington Post*, February 25, 2016 (https://www. washingtonpost.com/news/wonk/wp/2016/02/25/how-much-of-your-life-youre-wasting-on-your-commute/).

9. Abbey describes the "colorful history" of this essay in the preliminary remarks section of his essay collection *One Life at a Time, Please*.

10. Garrett Hardin, *Naked Emperor: Essays of a Taboo-Stalker* (Los Altos, CA: William Kaufman, Inc., 1982), 149.

Factors that Govern Birth Rates

Birth rates are primarily governed by four factors:

1. The number of children parents decide to have (which may not be the same as the number they would *like* to have if they had more money)
2. The age at which they decide to have children (that is, how long they postpone having their first child and how far apart they space any subsequent children)
3. The availability and affordability of contraception, sterilization, and abortion (so that parents do not exceed the number of children they've decided upon)
4. The proportion of the population that consists of women of reproductive age

The first three factors affect the number of babies born per woman, while the fourth factor is a function of the age structure of the population. Each of these factors is worth a closer look.

How many children people decide to have is influenced by cultural, economic, and personal factors. In the US, people now have far fewer children than their parents and grandparents did during the post-war baby boom (1946–64).[1] This is mainly because women have entered the workforce in large numbers. In fact, working women now outnumber working men in the US. Today's couples typically need two incomes to make payments on home mortgages, student loans, car loans, credit card debt, and to cover day-to-day expenses. Trying to combine paid work with child-rearing is challenging and stressful (a fact brought painfully home when the COVID-19 pandemic closed daycare centers and schools). The following table highlights some of the factors women have to consider:

Pros and Cons of Staying Home to Rear Children

Pros	Cons
Satisfaction from closer relationship with child. Don't miss out on events in child's life.	Less family income (a critical factor if the woman is single or if her partner does not make enough money to support the whole family).
Can breastfeed (better for the infant)	No safety net if partner loses job
More time to pursue own interests	Sacrifice the satisfactions of a career
Less stress	Earn less money when eventually return to work. Less retirement income.

When mothers return to work after a prolonged absence, they are generally paid less than their colleagues who did not interrupt their careers. This is because they missed out on regular pay raises and opportunities for advancement, and because their job skills may have become rusty. Employers may also assume that women with children will need to take more time off and therefore merit less pay. When prospective mothers weigh all the above factors, they increasingly opt to have few, if any, children.

How many children people decide to have is also strongly influenced by the cost of rearing them. In the United States, the government estimates that it costs about $250,000 to raise one middle-class child to the age of 18.[2] College expenses are on top of this. Given this high cost—made worse by America's 40-year decline in real income[3] and the fact that many would-be parents are still making payments on their college loans (which cannot be escaped by bankruptcy)—it is not surprising that families are becoming smaller and that large numbers of women are postponing childbearing until (they hope) they will be in a stronger financial position. This postponement reduces population growth in two ways: it lengthens the time between generations, and it makes conception more difficult (and often impossible), because fertility declines significantly with age.

Nowhere is the relationship between income and reproduction more glaring than in countries whose economies have severely weakened. In contemporary Greece, for example, high unemployment and underemployment are forcing many young Greeks to postpone marriage indefinitely. The Greek total fertility rate in 2021 was 1.35 children per woman (64 percent of replacement level). The supply of marriageable Greek men is shrinking, too, because many are migrating to more prosperous countries in quest of work.

The economic level below which people will refuse to breed varies according to culture and social class. This was pointed out by Malthus:

> In most countries, among the lower classes of people, there appears to be something like a standard of wretchedness, a point below which they will not continue to marry and propagate their species. This

standard is different in different countries, and is formed by various concurring circumstances of soil, climate, government, degree of knowledge, and civilization, &c. The principal differences which contribute to raise it are liberty, security of property, the spread of knowledge, and a taste for the conveniences and comforts of life. Those which contribute principally to lower it are despotism and ignorance.[4]

Another factor that influences family size is how much money parents hope to bequeath to their children when they die. The greater the number of children, the less wealth each child can inherit. This is true not only because the inheritance must be divided among a greater number of children, but because the parents of large families must spend much more of their lifetime earnings on rearing the children and therefore have less to pass on.

Before most women are willing to have children, they want to have a husband or partner who will share expenses and labor and provide emotional support. Finding such partners is getting harder and taking longer, which is another factor reducing the birth rate. One reason husbands are getting harder to find is a high male unemployment rate and men's diminishing earnings even when employed.[5]

When a woman succeeds in finding a partner, she has no assurance that the union will last. It may be terminated by abandonment, divorce, or death; or it may be temporarily interrupted by military deployment, imprisonment, or the need for the partner to move elsewhere to obtain work. Regardless of cause, women living alone are less likely to have babies than those who have a partner who can provide emotional support and contribute labor and money.

Ethnicity is another factor that influences how many children people decide to have. In 2020, the US birth rate per 1,000 women aged 15 to 44 was 62.8 for Hispanic women, 59.0 for black women, 53.2 for white women, and 50.1 for Asian women.[6] Were it not for the high Hispanic birth rate (which reflects the high immigration rate from Latin America), the US population would be declining (all other factors being equal). Ethnicity exerts its influence over family size through the values imparted to the young by parents and other respected figures. In cultures where having many children commands more respect than a good education and career, it is only natural that boys and girls aspire to have large families.

Religion also influences birth rates. Mormon culture, for example, has traditionally encouraged early marriage and large families. The state of Utah, which is 60 percent Mormon, has the highest birth rate in the United States. In 2016, the Utah fertility rate was 2.3 children per woman, compared to 1.8 for the nation as a whole.[7] Ultra-Orthodox Jews likewise produce large families. Depending on sect, they average six to eight children in the US. They mostly obtain the money to support these children by collecting rent as landlords. The Amish and Hutterites also have exceptionally large families. In the past, Roman Catholics also had large families, but American Catholics now use effective contraception at the same rate as Protestants. Demographer John Weeks suggested that religiosity retards the adoption of pro-contraceptive attitudes regardless of the specific content of the religious beliefs.[8]

Another factor that affects birth rates is the desire of parents to have a child of a particular sex. Some couples will keep producing babies until they get the sex they want (or finally give up). This high-lights the need for effective and affordable technologies to enable parents to have a baby of whatever sex they want on the first try.

The availability of contraception and sterilization is a very im-portant factor affecting birth rates. If approximately 70 percent of the women in a society use contraception, replacement-level fertility (about 2.1 children per woman in industrial societies and 2.3 chil-dren globally) will be achieved.[9] However, modern contraceptives are not a prerequisite for effective birth control. This was demonstrated by the steady decline in the US birth rate from 1850 through the 1930s—a decline brought about primarily because more people de-layed marriage or remained permanently celibate.[10] The term Mal-thus used for such prolonged celibacy was "moral restraint." It was one of three factors he believed could reduce the birth rate. The oth-er two were "vice" (abortion, infanticide, masturbation, and recourse to prostitutes) and "misery" (malnutrition, exhaustion, despair, and other harsh consequences of poverty).

In many countries, abortion plays at least as large a role in re-ducing birth rates as contraception. Some 40 million embryos and fetuses are aborted worldwide every year.[11] Demographers use two ways to express the frequency of abortion: the abortion *ratio* (num-ber of abortions per 1,000 live births) and the abortion *rate* (number of abortions per 1,000 women aged 15–44). The abortion ratio is a slightly better measure than the rate, because it relates abortion to actual, rather than potential, births. In 2017, the US abortion ratio was 18.4, while the abortion rate was 13.5 abortions per 1,000 women aged 15–44, down 22 percent from 16.9 in 2011. This is the lowest rate ever recorded in the United States. In 1973, the year abortion became legal, the rate was 16.3.[12] It is not known what percentage of living American women have had an abortion, but it can be projected from the current abortion rate that 24 percent of American women will have at least one abortion by the age of 45.[13]

Abortion rates in Latin America, Eastern Europe, Russia, and China are much higher than in the US. One reason for the high abortion rates in Catholic Latin America (despite the near-universal illegality of abortion there) is the Catholic Church's opposition to effective contraception and sex education, which contributes to a high rate of unwanted pregnancies. Western and Northern Europe have a low abortion rate, comparable to the US. Worldwide, about 44 percent of abortions are illegal and therefore endanger the mother's health and place her in legal jeopardy.[14]

Rape is another factor that can affect the birth rate. It increases the birth rate in societies where the morning-after pill and abortion are unavailable. The rapist may be a husband, boyfriend, casual date, or stranger. Rape is common during ethnic violence, as in Bosnia, Rwanda, the eastern Congo, and currently Ethiopia. The degree to which rape affects the birth rate in a particular country is unknown, but surveys suggest that a substantial percentage of teen pregnancies in the US are the result of rape by a date or boyfriend. In many cases, the males who perpetrate these rapes deliberately sabotage the girl's attempts at birth control.[15] Governments and religious institutions likewise sabotage birth control when they deny women access to sex education, contraception, and abortion. In Nicaragua, where all abortions are now banned, President Daniel Ortega prohibited a 12-year-old girl who had been raped by her stepfather from getting an abortion. The child was forced to carry the pregnancy to term and the baby was extracted by C-section. We shouldn't be surprised though: Ortega himself has been accused by his own stepdaughter of raping her over many years, starting when she was 11 years old.

Machismo, rape, denial of abortion, and "traditional family values" are closely related. All are rooted in the view that women are inferior to men and that men are therefore entitled to control women. These attitudes are deeply embedded in the abortion-prohibitionist religions, whose adherents believe that women should obey their husbands and are unworthy to serve as religious leaders (despite the fact that female religious leaders would be far less likely to sexually exploit children).

Another factor affecting birth rates is the proportion of the population that consists of women of reproductive age. This affects the birth rate independently of the number of children born per woman. Even after a population achieves replacement level fertility (about 2.1 children per woman), it will continue to grow for about 70 years (one lifetime) if there is a large proportion of young women in the population (which always happens after rapid population growth). In other words, even as the number of children per family declines, the total number of families increases, resulting in continued population growth. This phenomenon of continued growth even after replacement-level fertility has been achieved is called population momentum.

One factor known to heighten the desire of young women to have babies is the absence of a father when the girl is growing up. A father's absence also correlates with early menarche:

> Our findings suggest that father-absent girls reach menarche earlier and exhibit greater attraction to baby faces than father-present girls of the same age,

which may suggest greater readiness for parenting or a greater tendency to find opportunities to acquire parenting experience. In other words, by being more attracted to infant stimuli, rapidly maturing girls may acquire crucial parenting skills earlier in life and be better equipped for early reproduction and child-rearing.[18]

This indicates that keeping fathers present in families would help to lower birth rates by slowing the rate at which girls become sexually mature.

Before leaving the subject of the factors influencing birth rates, it is worth examining the factors behind America's postwar baby boom (1946–64). This long-lasting period of high fertility was mainly due to high demand for labor, which led to full employment and good wages. The factors behind the high labor demand were these:

- In the aftermath of World War II, there was high domestic and international demand for American manufactured goods
- America had no serious foreign competitors, because the economies of Europe and Japan had largely been destroyed
- There was high demand for young men in the labor force due to a low birth rate during the Depression and the deaths of nearly 300,000 American men during the war (and the disabling of many more)
- Women mostly worked as homemakers, and therefore did not compete with men for jobs

- Strict immigration quotas meant that immigrants were not major competitors for jobs
- Automation and computerization had not yet displaced large numbers of workers
- There were no international trade agreements to promote the offshoring of jobs
- A progressive income tax kept the nation's wealth from being excessively concentrated in the hands of a few.

All these factors combined to create full male employment and good incomes, making it possible for every man to marry and support a wife and children. But there will be no more American baby booms. Unemployment will inevitably rise due to computerization, robots, and the offshoring of jobs. And most of those lucky enough to keep jobs will see their wages decline. The rich will get richer, and the rest will get poorer. And if this trend is allowed to continue, America will eventually descend into violence, which will further decrease the birth rate and raise the death rate.

An interesting side effect of the American baby boom was its contribution to the breakdown of urban neighborhoods. A neighborhood can only remain stable as long as parents have an average of two children. When they have more than two, the excess children must leave the neighborhood when they grow up. If they fail to leave, the neighborhood's standard of living will fall as additional adults are crowded into houses that weren't built to accommodate them. When excess children move elsewhere, the rest of the family often follows. In the United States, during the latter half of the 20th century, there was a dual migration of urban whites into the suburbs and inner-city blacks into the former white neighborhoods. The baby boom was not the only factor behind these twin migrations (urban renewal projects, for example, forced many blacks to migrate), but it made them inevitable. Both races ended up losing their old neighborhoods where everyone was known.

Notes

1. In 1960, the US birth rate was 118 per 1,000 women aged 15–44. By 1980, it had fallen to 68, where it has more or less remained. See http://www. childtrendsdatabank.org/pdf/79_PDF.pdf.

2. Mark Lino, "Expenditures on Children by Families, 2009," US Dept. of Agriculture, Center for Nutrition Policy and Promotion, Miscellaneous Publication No. 1528-2009 (2010).

3. According to MIT economist Michael Greenstone, real median wages for men in the US have fallen 28 percent (about $13,000) since their peak in 1972. This figure is the median for *all* men, not just full-time workers. It therefore takes into account the many men who have been pushed out of the workforce. See also David Leonhardt, "The Struggles of Men," *New York Times*, March 4, 2011. The minimum wage in the US has declined 20 percent since 1971, adjusted for inflation. Source: Eric Schlosser, "The Food Movement: Its Power and Possibilities," *The Nation*, October 3, 2011, p. 18. In his 1998 article "A Rising Tide Is Not Lifting All Ships," Albert Bartlett compares profits/economy, wages/ economy, and profits/wages in the US. He found that profits were growing at 25 times the rate of wages. His article is included in *The Essential Exponential! For the Future of Our Planet* (see bibliography).

4. Thomas Robert Malthus, *An Essay of the Principle of Population*, book 4, chapter 9 (1798).

5. See "What Me, Marry?" by Kate Bolick in *The Atlantic*, November 2011.

6. See https://www.cdc.gov/nchs/data/vsrr/vsrr012-508.pdf.

7. Interestingly, if the Mormon Church had retained its original practice of polygamy, the birth rate would be lower. That's because a male polygamist must divide his sexual attentions among two or more women. As a result, each wife receives his semen less frequently than she would in a monogamous marriage. This reduces the birth rate for each wife. Depending on the number of wives, polygamous marriages deny an equivalent number of men the opportunity to marry. These wifeless men often turn to prostitutes, and few offspring result from such unions.

8. John R. Weeks, *Population: An Introduction to Concepts and Issues,* 6th ed. (Wadsworth Publishing Company, 1996), 324.

9. Ibid., 115.

10. Ibid., 155.

11. See http://www.guttmacher.org/in-the-know/abortion.html.

12. The best source of information on abortion rates in the United States and around the world is the Guttmacher Institute (guttmacher.org).

13. See http://www.guttmacher.org/media/presskits/abortion-US/index.html.

14. Ibid.

15. Roni Caryn Rabin, "Report Details Sabotage of Birth Control," *New York Times*, February 15, 2011, http://www.nytimes.com/2011/02/15/health/ research/15pregnant.html.

16. See https://www.macrotrends.net/countries/CHN/china/fertility-rate.

17. See www.apa.org/science/about/psa/2004/01/maestripieri.

Immigration's Role in Fueling Population Growth

Most immigrants—at least 90 percent by most estimates—migrate in order to improve their economic status. The remaining 10 percent migrate to obtain more freedom or escape persecution.[1] Yet many who migrate for freedom and safety also hope to improve their economic status. So the two categories are not sharply distinct.

From a census standpoint, emigration is like death, while immigration is like birth. But unlike real death and birth, emigration and immigration are reversible: immigrants can, and often do, return home—either voluntarily or forcibly. Because immigration consists of subtracting an individual from one country and adding him or her to another, we might think that this is a neutral process that does not cause a net increase in the world's population. But the reality is that people who migrate leave behind a vacant habitable niche (albeit a poor one). This niche is nearly always promptly filled by further reproduction. The population of Mexico, for example, is far larger than it was 35 years ago when mass migration to the United States began. Malthus was aware of this paradoxical phenomenon, pointing out that "it has been particularly remarked that the two Spanish provinces from which the greatest number of people immigrated to America became in consequence more populous."[2] Part of the explanation may be the remittances sent home by immigrants.

Overpopulation, by generating poverty, hunger, war, crime, and environmental degradation, has always been the prime driver of emigration.[3] Today it is driving Mexicans to the US, Albanians to Greece, Zimbabweans to South Africa, North Koreans to China, Indonesians and Filipinos to the Arabian Peninsula, North Africans and Middle Easterners to Europe, and Chinese and Indians to all points. About 60 percent of global migrants head for industrialized countries, but migration from one developing country to another is also increasing rapidly.[4] In 2013, the total number of international migrants was estimated to be 232 million. The United States has 5 percent of the world's population, but 20 percent of the world's migrants, and over 25 percent of them are in the US illegally.[5] Europe has about 10 percent of the world's population and about 30 percent of the world's immigrants.[6] In the year 2015 alone, Germany's net migration was 1.14 million. Mass migration is also occurring *within* countries as people flee the poverty of rural life for the squalor and exploitation (but sometimes greater economic opportunities) of the big city. The ratio of internal migrants to international migrants is 4:1.[7] The flood of migrants will increase greatly in coming decades. It is estimated that by 2035 Africa will have as many people of working age as all the rest of the world combined.[8] Most of them will be unemployed and hungry, which means they will try to migrate out of Africa by the tens of millions. They will follow in the footsteps of our hominid ancestors who also left Africa because they overpopulated.

An ugly feature of immigration is the violence it triggers when immigrants clash with natives over resources. Well-known examples are the wars between American Indians and European settlers, and the enduring conflict in Palestine between Jewish settlers and indigenous Arabs. These conflicts are always characterized by racism and not infrequently lead to massacres and genocide. Sometimes the immigrants prevail (as they have in the United States, Australia, and—for the time being—Israel), and sometimes the natives prevail (as in the Dominican Republic, where the Dominican army slaughtered some 20,000 Haitian immigrants in 1937). In the United States, immigrants have mostly been tolerated because the economy has generally been strong, and, until recently, most immigrants shared the same cultural background as most natives. Immigrants from other backgrounds, however, were subjected to discrimination. During World War II, for example, US citizens of Japanese descent were rounded up and interned for the duration of the war. In contrast, American citizens of German and Italian descent were left alone (although some German and Italian nationals were interned).

The discussion that follows will focus on immigration as it affects the United States. Unless otherwise indicated, the statistics presented here were obtained from the websites of the US Census Bureau and the Pew Research Center.

The United States and Europe are the preferred destinations of most of the world's immigrants. Annually, the US acquires slightly more than 1 million legal immigrants and roughly 0.5 million illegal immigrants.[9] So the annual total is about 1.5 million. About half of the illegal immigrants cross the border illegally, while the other half cross legally, but then become illegal when they overstay their visas or Border Crossing Cards.

Every year, the US issues hundreds of thousands of permanent immigrant visas ("green cards"). These visas allow immigrants to permanently live and work in the US (and reproduce). There are no limits to the number of green cards that can be issued to immediate relatives of US citizens (defined as spouses, parents, and children under age 21). Once these family members arrive in the US and become lawful permanent residents, they can legally work in the US. After they've lived in the US for 5 years, they can apply for US citizenship.

In addition to immediate family members (spouses, parents, and children under 21), the siblings and adult children of US citizens can also be admitted. So can the spouses and unmarried children (whether young or adult) of *noncitizens* who are lawful permanent residents. These admissions of nonimmediate relatives are called "family preference" admissions. Unlike the unlimited admissions for immediate relatives, family preference admissions have annual limits. By law, the minimum to be admitted is 226,000 and the maximum is 480,000 minus the number of visas issued to immediate relatives and parolees (a minor category), plus unused employment visas from the previous fiscal year. The following table shows the 2021 distribution of family preference quotas:

Quotas allotted to US citizens	Unmarried adult children	23,400*
	Married adult children	23,400**
	Brothers and sisters	65,000***
Quotas allotted to lawful permanent residents	Spouses and minor children	87,900
	Unmarried adult children	26,300

*Plus any unused visas from the 3rd row.
**Plus any unused visas from the 1st, 4th, and 5th rows.
***Plus any unused visas from all the other family preferences.

When the number of immediate family visas is added to the number of family preference visas, the total often exceeds 480,000 per year.

Yet another visa is the diversity visa, which grants admission to up to 55,000 people from countries that are historically underrepresented in US immigration. The winners of these diversity visas are drawn randomly from a pool of 10 to 12 million applicants.

There is also an EB-5 visa that allows wealthy foreigners and their immediate family members to buy their way into the US by investing at least one-half million dollars into the US economy. About 90 percent of the thousands of EB-5 visas issued each year go to Chinese businessmen and their families, including corrupt officials. Conveniently for these officials the US has no extradition treaty with China.[10] Although the intent of the EB-5 plan was to provide additional stimulus to the US economy and provide jobs in areas of high unemployment, most of the investments have instead been made in the wealthy downtowns of Manhattan, Miami, Las Vegas, Los Angeles, and other large cities. The program is poorly supervised and has enriched real estate developers.

Each year, the US also grants many temporary "nonimmigrant" visas to tourists, foreign students, and workers (both skilled and unskilled). Although there are limits on the number of visas issued to workers, there are no limits on the number issued to tourists and foreign students. There is also a temporary visa administered by the State Department, which allows more than 100,000 foreign college students to be admitted to the US each year to work during their school vacations.

The US also admits many refugees each year. The annual number admitted is determined by the president in consultation with Congress. The annual refugee limit for fiscal year 2021 was 62,500. This number is divided among the world's geographical regions as follows: 22,000 from Africa, 6,000 from East Asia, 4,000 from Europe and Central Asia, 5,000 from Latin America and the Caribbe-an, 13,000 from the Near East and South Asia, plus an unallocated reserve of 12,500.

Those who are interested in learning more about how the US allocates visas can consult the website of the American Immigration Council (www.americanimmigrationcouncil.org).

US Immigration Fuels Rapid Population Growth

The rapid rise in US immigration since 1965 has been the princi-pal contributor to US population growth. Between 1965 and 2015, immigrants and their descendants added 72 million people to the US population. If immigration continues at its present rate, it will expand the US population by an additional 103 million by 2065.[11]

In 1970, only 1 in 21 American residents was foreign born (4.7 percent of the population), but by 2021 the figure had climbed to more than 1 in 7 (14.4 percent).[12] For comparison, the black popu-

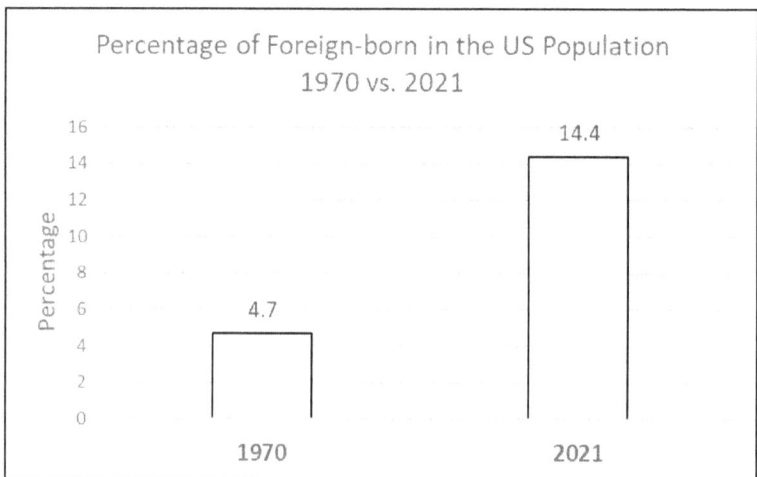

Percentage of Foreign-born in the US Population 1970 vs. 2021

lation of the US is only 12 percent. Immigrants and their US-born children now comprise 26 percent of the overall US population.

Half of US Immigrants (Legal and Illegal) Come from Latin America

About 28 percent of the foreign-born in the US are natives of Mexico.[13] In fact, 10 percent of all the people born in Mexico who are alive today now live in the United States. In addition to Mexico's contribution, 20 percent of US immigrants come from other Latin American nations, especially Cuba, Guatemala, Honduras, El Salvador, and, most recently, Venezuela and Nicaragua. Altogether, Latin Americans comprise slightly more than 50 percent of the foreign-born in the US.

Immigration's Negative Impact on the Environment

When people migrate from a poor nation to a rich one, they greatly increase their consumption of resources and their output of pollutants. If we compare the amount of CO_2 pollution produced by an average citizen of China to that produced by an average American, we find that each Chinese in 2010 produced 6.05×10^{-3} tons of CO_2, while each American in that same year produced an average of 17.6×10^{-3} tons of CO_2.[14] So when Chinese immigrate to the US, they begin to release three times more CO_2 than they would have in China. And their children will likewise pollute at three times the Chinese rate. Mass migration from poor countries to rich countries causes immense environmental harm by increasing the rate at which resources are consumed and the rate at which pollution is generated.

The homesickness of immigrants also contributes to global pollution. Immigrants naturally miss their friends and relatives and native culture, so they travel home as often as they can. This results in millions of long-distance flights that consume nonrenewable fuel and produce pollution.

Incompatibility of Some Immigrant Values with Western Values

Most US immigrants come from authoritarian states where obedience and credulity are rewarded, while free thought and free action are discouraged and often severely punished. Authoritarian states nearly always have cultures in which graft, bribery, nepotism, and tax evasion are pervasive and tolerated. In a psychological experiment in which 2,500 subjects from 23 countries rolled dice and then announced the results for a reward, it was found that "higher rates of corruption, tax evasion, and political fraud in a subject's country predicted higher rates of lying. This is no surprise…high rates of rule violations in a community decrease social capital, which then fuels individual antisocial behavior."[16]

Thomas Jefferson worried about the impact on American culture and legislation of immigrants from authoritarian states:

> Every species of government has its specific principles. Ours perhaps are more peculiar than those of any other in the universe. It is a composition of the freest principles of the English constitution, with others derived from natural right and natural reason. To these nothing can be more opposed than the

maxims of absolute monarchies. Yet, from such, we are to expect the greatest number of emigrants. They will bring with them the principles of the governments they leave, imbibed in their early youth; or, if able to throw them off, it will be in exchange for an unbounded licentiousness, passing, as is usual, from one extreme to the other. It would be a miracle were they to stop precisely at the point of temperate liberty. These principles, with their language, they will transmit to their children. In proportion to their numbers, they will share with us the legislation. They will infuse into it their spirit, warp and bias its direction, and render it a heterogeneous, incoherent, distracted mass. I may appeal to experience, during the present contest, for a verification of these conjectures. But, if they be not certain in event, are they not possible, are they not probable?[17]

Illegal US Immigrants

The Pew Research Center estimated that in 2017 there were 10.5 million illegal immigrants living in the US (3.5 percent of the 2017 US population).[18] This represents a decline from a peak of about 12.2 million in 2007. Of the 10.5 million illegal immigrants, about 8 million are in the labor force (defined as either working or looking for work).[19] This represents 5 percent of the US labor force—about the same as the official unemployment rate.

Although illegal immigrants make up 3.3 percent of the US population, their children comprise 7 percent of all US children due to the higher birth rates of illegal immigrants. In 2016, 6 percent of all US births were to illegal immigrants. This was down from 9 percent in 2007.[20]

How Illegal Immigrants Gain Legal Status

Many illegal US immigrants have subsequently been able to achieve legal status. One way they have done this is through a law that allows American-born children of illegal immigrants to sponsor their parents for legal status once the children turn 21. US law has permitted this since 1965, and 79 percent of illegal immigrant children in the US, once they turned 21, have taken advantage of this law to serve their parents as "anchors" for obtaining legal status. There are no quotas on how many foreign parents can be legalized this way.

Another way illegal immigrants have obtained legal status is through government amnesties. In 1986, a Reagan administration amnesty enabled 2.9 million illegal immigrants to obtain legal status. In the 1990s, a series of amnesties by the Clinton administration enabled another 3 million illegal immigrants to obtain legal status. In 2012, the Obama administration announced the Deferred Action for Childhood Arrivals (DACA) program. This program allows those who were less than 31 years old when the rule went into effect to apply for deportation deferral and obtain a work permit if they entered the country before June 15, 2007. In 2014, President Obama issued an executive order that expanded DACA by granting deportation relief and temporary legal status to anyone who entered the US before they were 16 and arrived before January 1, 2010, provided they either had the equivalent of a high school degree (or were working on it) or were honorably discharged from the military. This expansion of DACA also extended the work permit from 2 years to 3 years.

Economic Burden of Immigrants

Most US immigrants are not as well educated as natives. In fact, about 31 percent of today's adult immigrants have not completed high school, compared to 8 percent of natives.[21] As a consequence of their lower educational status and the challenge of mastering English, immigrants are more likely than natives to be poor, lack health insurance, and receive welfare services. The poverty rate for immigrants and their US-born children is 17 percent, nearly 50 percent higher than that of natives.[22] The percentage of immigrants lacking health insurance is 34 percent, compared to 13 percent of natives.[23] Immigrants and their US-born children account for 71 percent of the increase in the medically uninsured since 1989.[24] Immigrant children account for nearly all the increase in US public school enrollment over the last 20 years.[25] In 2009, the percentage of immigrant-headed households with children under 18 using at least one welfare program was 57 percent, versus 39 percent for native

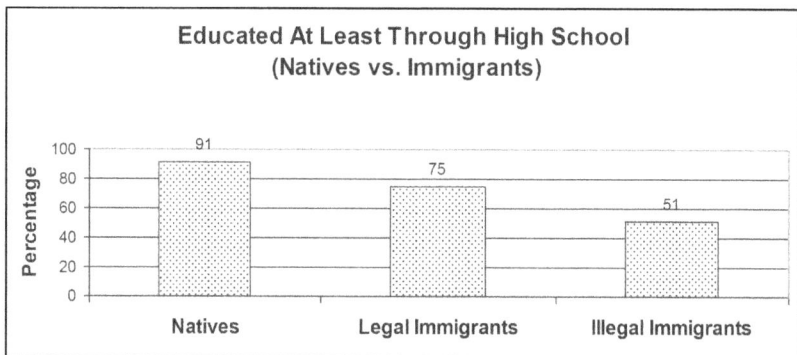

Educated At Least Through High School
(Natives vs. Immigrants)

Natives	91
Legal Immigrants	75
Illegal Immigrants	51

households with children.[26] Although immigrants do improve their economic status over time, even after 20 years they are still poorer on average than natives.

Lautenberg Amendment

In 1989, the US Congress passed the Lautenberg Amendment (Public Law 101-167), which permitted people from the former Soviet Union who were either Jewish or minority Christians (mostly evangelicals) to immigrate to the United States as refugees. This bill also allowed certain categories of people from Indochina and Iran to come to the US. Unlike ordinary refugees, who were required to establish "a well-founded fear of persecution on a case-by-case basis," the groups covered by the Lautenberg Amendment only needed to demonstrate "a credible, but not necessarily individual, fear of persecution." In practice, this meant they merely had to declare a fear of persecution. Also, unlike ordinary immigrants, the Lautenberg immigrants were immediately eligible for government assistance programs including Medicaid, food stamps (SNAP), special cash assistance, and Temporary Assistance for Needy Families.

Not surprisingly—for that was always its intent—the Lautenberg Amendment enticed nearly 400,000 Jews from the former Soviet Union to immigrate to the US. By 2010, the total number of Lautenberg immigrants, including Indo-Chinese and minority Christians, was 440,000. Many of these immigrants were elderly and were therefore placed immediately on welfare. In the decades since, American taxpayers have supplied them with basic income, apartments, food, and medical care. And whenever social workers deem they would benefit from home care, they are assigned a caregiver for a certain number of hours per week. These caregivers clean their apartments, cook their meals, translate, take them shopping, and transport them to doctor appointments, the grocery store, and the pharmacy. Some of these Lautenberg immigrants have figured out that they can generate additional cash by having their caregiver work only a portion of the allotted hours, kicking back the unworked hours as cash when the caregiver is paid. Another way Lautenberg immigrants generate cash by giving their food stamp card to their caregiver and demanding that the caregiver pay them back the value of the card. Some of them also designate a son or daughter as their caregiver, thereby keeping the money in the family. Even when Lautenberg immigrants live with wealthy sons and daughters, they remain eligible for food stamps, caregivers, and other amenities.

The Lautenberg Amendment was premised on the claim that Soviet Jews were under threat of persecution. In reality, Soviet Jews were under no threat at the time the Lautenberg Amendment was enacted (nor at any time since). Their preferential admission to the US meant that many people who were truly in need of refuge were excluded. Senator Lautenberg, who was the son of Ukrainian Jewish immigrants, privileged his own ethnic group at the expense of those who were truly in need of refuge, and at the expense of American taxpayers. Yet the US Congress, under pressure from the powerful Jewish lobby, regularly renews the Lautenberg Amendment.

Hardships Endured by Immigrants and Their Children

Immigration is hard on immigrants. The trouble begins when families are split apart as some members migrate and others remain behind. This family disunification becomes especially challenging when a relative who remained behind becomes seriously ill. The immigrant must then decide whether to quit work and return home or remain in the adopted country to continue providing for his or her family. Family separation also occurs involuntarily when illegal immigrants are arrested and deported, leaving their spouses and children to fend for themselves.

Immigrants also face hardships in adjusting to unfamiliar customs and learning a new language. Not surprisingly, many choose to live in ethnic enclaves where they can continue to practice their traditional culture and speak their native tongue.

Immigration also creates conflict between immigrant parents and their children. The parents naturally adhere to the values they were taught in youth, while their children seek to blend in with their peers. In some immigrant cultures, these intergenerational conflicts have led to so-called honor killings, in which young women are murdered by male family members for such acts as refusing an arranged marriage, demanding a divorce, secretly seeing a man, refusing to wear a hijab, or even for having been raped.

Immigrants may also suffer discrimination from natives who fear job competition or may dislike the immigrant's race, dress, or religion. And when immigrants feel rejected and see no path forward, they may turn to crime or even terrorism. In short, immigration is much harder on immigrants than they were led to believe. That is why so many of them eventually return home. And more would do so if their children had not put down roots in the new land.

Many immigrants also suffer during their journeys to the new land. Some drown while trying to cross the sea in flimsy watercraft or die in fiery crashes of smugglers' trucks. Some perish of dehydration in the desert or freeze to death while trying to enter the US from Canada (as recently happened to a family from India). And many immigrants suffer robberies and rapes during their migrations.

Because immigration is hard on immigrants and contributes to harmful population growth while lowering wages and employment for our most vulnerable citizens, every effort should be made to convince people to stay in their own country. Wealthy nations need to reduce their quotas for legal migration and strengthen barriers against illegal immigration. They also need to establish programs to educate the world's poor about the harm of overpopulation and ensure that they have access to long-term contraception, the morning-after pill, and safe abortion. Moreover, we need leaders who will not be afraid to condemn religious doctrines that oppose contraception and abortion. If we took these measures, we could prevent the creation of both legal and illegal immigrants and ensure a better life for all.

Immigration's Role in Increasing Unemployment and Depressing Wages

It is often claimed that Americans would never consent to perform the kinds of jobs that many immigrants—especially illegal immigrants—typically perform. From this it is argued that immigrants do not compete against natives for jobs. The problem with this argument is that its premise is false. The jobs now performed by immigrants were once performed by Americans, and Americans would likely take these jobs again if wages and working conditions were sufficiently improved. But even if Americans refused to take these jobs, it wouldn't matter: we would simply mechanize the jobs, or find substitutes for the products and services, or learn to live without them.

It is mainly the less well-educated Americans who are forced to compete against the foreign born for jobs and who therefore suffer most from the current mass-immigration policy. Economist Dean Baker explains:

> Immigration policy has also been structured and enforced in a way that widens income gaps. Specifically, the lax enforcement of immigration laws amounts to an implicit policy of allowing undocumented immigrants to work in low-paying jobs. By increasing the supply of low-wage labor, this policy drives down wages for native-born workers who might otherwise hold these jobs.[27]

No class of Americans has suffered more from US mass immigration policy over the past 120 years than the descendants of African slaves. Successive waves of immigrants have flooded the US labor market, preventing blacks from getting good jobs and pushing them out of the occupations they once dominated. Their only respite was the period from 1924 to 1965 when Congress reduced immigration to moderate levels. That 41-year period of low immigration greatly increased demand for labor, spurring the Great Migration of blacks from the rural South to the industrial North. This led to the percentage of blacks in the middle class rising from 22 percent in 1940 to 71 percent in 1980.[28] This economic improvement helped spur the civil rights and voting rights movements of the 1960s. But when mass immigration resumed following the passage of the 1965 Hart-Cellar Immigration and Nationality Act, blacks who lacked a college education suffered a 25 percent drop in employment and a massive decline in wages.[29] One consequence of the high unemployment rate among black males is that two-thirds of black families are now headed by a single woman—a circumstance that makes escaping poverty much more difficult. In 2016, the median net worth of an American white family was $171,000, while the median net worth of a black family was $17,150—a ratio of nearly 10 to 1.[30] In his book *Back of the Hiring Line*, journalist Roy Beck documents the sorry history of immigration's destructive impact on America's black population.

It is not only working-class Americans who are harmed by mass immigration: many Americans in the STEM occupations (science, technology, engineering, and math) are also being excluded from jobs. This is notably true in Silicon Valley, where CEOs claim they can't find qualified Americans to do the work. These CEOs petition the government to admit more foreign software engineers, because they can pay the foreign workers less.

Many recent graduates of US medical schools are also harmed by current immigration policies. Every year, thousands of these graduates are denied residencies in teaching hospitals because of preferences given to foreign graduates. In 2021, more than 4,000 foreigners received medical residencies despite the fact that there are at least 7,000 US citizens and lawful residents who have been denied residencies. Without a residency, these medical school graduates cannot practice medicine or even begin to pay off their educational debts (which average over $200,000). Moreover, the granting of residencies to thousands of foreigners means that their native lands (where they were educated at great public expense) are denied the benefit of their badly needed skills.[31]

The death rate for middle-aged white American males has been rising 0.5 percent annually for the last decade.[32] The official causes for this rising mortality are suicide and accidental death from alcohol and opiates. Much of the hopelessness that leads to suicide and addiction is due to unemployment and falling wages, which have left millions of Americans less well off than their parents. Mass immigration is a major contributor to this tragedy.

It is not only adults who suffer from the current US immigration policy. America's teenagers now find it much harder to obtain summer employment and part-time jobs because of competition from immigrants. This problem has been exacerbated by the State Department's Summer Work Travel (SWT) program, which brings more than 100,000 young foreign college students to work in the US during their summer vacations.[33] Competition from immigrants is hard on America's youth and has negative consequences for society. Many young people who end up in prison would not be there if they had been integrated into the economy instead of being excluded. Suicide is now the third leading cause of death for Americans 10 to 24 years old, and the suicide rate has been rising since 1999.[34]

Our present immigration policy benefits private interests at the expense of the public. Businessmen benefit from having more consumers for their products and services, and they reap the profits of cheap immigrant labor, while the public picks up the costs. These costs include medical care, welfare programs, education (including bilingual education), crime, lower wages for natives, more unemployment (including lack of jobs for teenagers), and a less cohesive society.

The elites in the immigrant-exporting countries also benefit from the present immigration policy. It relieves them of large numbers of young people who would otherwise create trouble by demanding jobs and justice. And they know that these young people will work hard in their new country, sending home remittances to their families—remittances that play a vital role in propping up the local economy.[35] In fact, the sum of the world's remittances is now four times greater than all international development aid.[36]

Former World Bank economist, Herman Daly, commented on how the interests of America's working class are being sacrificed to current policies on immigration and trade:

> It is considered impolite to talk about the political interest in cheap labor in this country, or about the use by the employer class of free trade and unenforced immigration laws as instruments for promoting lower wages and higher profits. Liberal intellectuals, who one might have hoped would see through the mystification, instead have advocated free trade and easy immigration as a way of being generous at someone else's expense, and of proving to themselves once again that they are not racists or even nationalists. And the economists assure them that economic growth will eventually make everyone better off, so whatever temporary cost may fall on our laboring class is a small price for "us" to pay for the gratifying illusion that we are helping starving people across the sea.[37]

The Role of the Mass Media in Promoting Mass Immigration

Because mass immigration makes the rich richer, it is only natural that they use the mass media (which they own) to proclaim a positive message about immigration. In print, radio, television, websites, and social media, the big media companies tell us over and over that immigrants are ambitious, inventive, industrious, entrepreneurial, and are the salvation of our inner cities. Even if all this lavish praise were true, it has a dark side, for it falsely implies that our native population is ignorant, lazy, and contemptible.

The mainstream media also tell us that we should welcome mass immigration because it broadens our cultural horizons, exposing us to new ideas, new foods, new music, and new fashions. That is certainly true, but not everything novel is desirable. Most of us would prefer to do without such novelties as fatwas, *narcocorridos*, female circumcision, niqabs, honor killings, warring gangs, and terrorism. A far better way to expand our cultural horizons would be to learn foreign languages, read books, and travel (both physically and on the internet).

> Useful diversity is more efficiently attained by transporting images, ideas, and dreams between geographically fixed populations rather than uprooting and moving human bodies. Pure information can be

moved more cheaply than information wrapped in human bodies.

—Garrett Hardin[38]

Journalist and educator Richard Heinberg highlighted some of the political and business interests behind current US immigration policy:

> Every survey since the 1940s has shown that a majority of Americans favors reducing immigration, yet during that time legal immigration has quadrupled (it doubled during Bush I and again during Bush II). Much of the support for liberalizing immigration policy has come from the Democratic Party (in its calculus, more immigrants means more Democrats), as well as from the construction industry (more immigrants equal more housing starts), the food industry (which depends on low-paid seasonal farm workers), and the US Chamber of Commerce (immigrants reduce labor costs).[39]

Opponents of mass immigration are sometimes accused of being racists and bigots. No doubt some are. But the plan presented in this book is fair to everyone. Any foreigner who acquires two credits by donation or purchase would be allowed to immigrate, regardless of nationality, religion, sexual orientation, skin color, age, level of education, or political beliefs (provided they don't advocate criminal acts and are willing to abide by the US Constitution).

Countries that adopt large numbers of immigrants (like the United States, Germany, Italy, and the United Kingdom) need to face the fact that mass immigration is, on balance, detrimental. Beyond that, every country needs to determine an optimal size for its population and take the necessary measures to achieve it. We expect couples to stay within their means when they produce children or adopt, and we should demand no less of nations—especially our own.

Notes

1. Philip Martin, "The Global Challenge of Managing Migration," prb.org.

2. Thomas Robert Malthus, *An Essay on the Principle of Population* (1798), book 2, chapter 13.

3. An interesting example of how overpopulation triggers migration is provided by an aphid *(Aphis nerii)* which feeds and multiplies on milkweed plants. As soon as these wingless aphids overpopulate a plant, they sprout wings so they can fly off to find new milkweed plants. When conditions on a milkweed plant are especially crowded, the young are even born with wings! Outstanding photos showing the life history of these aphids can be seen at www.zenthroughalens. com. When you get to this website, just search for "aphids."

4. Philip Martin, "The Global Challenge of Managing Migration," prb.org.

5. Ibid.

6. Ibid.

7. Ibid.

8. See https://www.bloomberg.com/news/articles/2015-04-28/africa-s-labor- force-newcomers-to-exceed-world-by-2035-imf-says.

9. The Department of Health and Human Services website at http:// www.dhs.gov/xlibrary/assets/statistics/publications/lpr_fr_2009.

pdf states that 1,130,818 immigrants were granted permanent legal resident status in 2009. Illegal immigration may have fallen to about 500,000 in 2011, due to better border enforcement, fewer job opportunities in the US, and more job opportunities in Mexico. However, since 2014, it has been rising again, as illegal immigrants have learned to take advantage of America's lax asylum policy.

10. Anthony Kuhn, "When Corrupt Chinese Officials Flee, The US Is a Top Destination," National Public Radio, June 23, 2015. Can be viewed at http://www.npr.org/sections/ parallels/2015/06/23/416828057/when-corrupt-chinese-officials-flee-the-u-s-is-a-top-destination.

11. See http://www.pewhispanic.org/2015/09/28/modern-immigration-wave-brings-59-million-to-u-s-driving-population-growth-and-change-through-2065/ph_2015-09-28_immigration-through-2065-02/.

12. See http://www.migrationpolicy.org/article/frequently-requested-statistics-immigrants-and-immigration-united-states.

13. See http://www.migrationpolicy.org/article/mexican-immigrants-unit- ed-states. Also see "Mexican Immigrants in the United States, 2008," Pew Hispanic Center, April 15, 2009.

14. This is a straightforward calculation: simply take the estimated total production of CO_2 for all of China (8,286,892 tons in 2010) and divide by the 2010 Chinese population (1,370,536,875). For the US, the 2010 production of CO_2 was 5,433,057 tons, and the 2010 population was 308 million.

15. Michael Lipka, "Muslims and Islam: Key Findings in the US and around the World," December 7, 2015 (www.pewresearch.org/ fact-tank/2017/08/09/muslims-and-islam-key-findings-in-the-u-s-and-around-the-world/).

16. *Behave: The Biology of Humans at Our Best and Worst*, p. 514. The essence of social capital is shared values and trust. It is what enables a society to function effectively.

17. Thomas Jefferson, *Notes on the State of Virginia*, query VIII, 211.

18. See www.pewresearch.org/fact-tank/2016/11/03/5-facts-about-illegal-immigration-in-the-u-s/.

19. Ibid.

20. See www.pewresearch.org/fact-tank/2018/11/01/the-number-of-u-s-born-babies-with-unauthorized-immigrant-parents-has-fallen-since-2007/.

21. Steven A. Camarota, "Immigrants in the United States, 2007: A Profile of America's Foreign-Born Population," Center for Immigration Studies, November 2007. Can be viewed at http://www.cis.org/immigrants_profile_2007.

22. See http://pewhispanic.org/files/factsheets/47.pdf and http://pewresearch.org/pubs/1876/unauthorized-immigrant-population-united-states-nation- al-state-trends-2010.

23. Ibid.

24. Ibid.

25. Steven A. Camarota, "Immigrants in the United States, 2007: A Profile of America's Foreign-Born Population," Center for Immigration Studies, November 2007. See http://www.cis.org/immigrants_profile_2007.

26. Steven A. Camarota, "Welfare Use by Immigrant Households with Children: A Look at Cash, Medicaid, Housing, and Food Programs," Center for Immigration Studies, April 2011.

27. See http://cis.org/Reprinted with permission of the publisher. From *Plunder and Blunder* copyright © 2008 by Dean Baker, page 16.

28. James P. Smith and Finis R. Welch, "Black Economic Progress After Myrdal," *Journal of Economic Literature* 27, no. 2 (June 1989): pp. 519–564.

29. George J. Borjas, Jeffrey Grogger, and Gordon H. Hanso, "Immigration and African-American Employment Opportunities: The Response of Wages, Employment, and Incarceration to Labor Supply Shocks," NBER Working Paper no. 12518 (September 2006).

30. See https://www.brookings.edu/blog/up-front/2020/02/27/examining-the-black-white-wealth-gap/.

31. See Emma Goldberg, "I Am Worth It: Why Thousands of Doctors in America Can't Get a Job," *New York Times*, February 19, 2021 (www.nytimes.com/2021/02/19/health/medical-school-residency-doctors.html). The shortage of US doctors is due to 5 factors: (1) a growing population (due to mass immigration), (2) an aging boomer population with deteriorating health, (3) an aging doctor population (many of whom will soon retire), (4) the failure of US teaching hospitals to match all US medical graduates to residencies, and (5) the stress of the COVID-19 pandemic, which has prompted many doctors to retire early.

32. Rob Stein, "In Reversal, Death Rates Rise for Middle-Aged Whites," National Public Radio, November 2, 2015. The research was conducted by Nobel laureate and professor of economics Angus Deaton and his wife, Ann Case, both of Princeton University. Their findings were published in *Proceedings of the National Academy of Sciences*.

33. The SWT program gives employers a strong incentive to hire foreign students rather than Americans, because employers don't have to pay taxes for Social Security, Medicare, and unemployment compensation. This saves employers about 8 percent over hiring

Americans. Although the State Department calls SWT a foreign exchange program, it is really a program of cheap labor that undermines American employment.

34. See https://www.cdc.gov/violenceprevention/suicide/youth_suicide.html.

35. In 2014, the Government Accountability Office estimated that $54 billion was sent from the US to other countries. Of this, $25 billion went to Mexico. That was more money than Mexico made from oil production. In Lebanon, remittances make up a remarkable 22 percent of GDP.

36. Webinar by Philip Martin, January 2014, prb.org.

37. Herman Daly, *Beyond Growth: The Economics of Sustainable Development* (Boston: Beacon Press, 1996), 156. Another problem with mass immigration is that it reduces worker productivity. Any time the supply of labor is plentiful, employers lack an incentive to invest in labor-saving technology or invest in better educating their workforce. This results in lower per capita productivity.

38. Garrett Hardin, *Living within Limits* (New York: Oxford University Press, 1993), 277.

39. Richard Heinberg, *The End of Growth* (British Columbia: New Society Publishers, 2011), 214.

Tragedy of the Unregulated Commons

The tragedy of the unregulated commons is the exhaustion of resources that occurs when resources are available to everyone at little or no cost. It was first described by ecologist Garrett Hardin in a 1968 paper published in *Science*.[1] In order to prevent the tragedy of the commons, we must regulate the consumption of resources that we hold in common. John Stuart Mill emphasized the importance of such regulation:

> These are the inheritance of the human race, and there must be regulations for the common enjoyment of them. What rights, and under what conditions, a person shall be allowed to exercise over any portion of his common inheritance, cannot be left undecided. No function of government is less optional than the regulation of these things, or more completely involved in the idea of a civilized society.[2]

In recent decades, mankind has finally begun to regulate air and water pollution, ration water in arid regions, ban random dumping of waste, and regulate soil erosion, logging, hunting, fishing, and grazing on public land and water. But if populations continue to grow, no amount of regulation can prevent further environmental deterioration. Every child added to a population imposes costs on all the other members by reducing shared resources and increasing pollution. Yet the parents do not have to compensate the other members for this resource loss and pollution, any more than a new fisherman has to compensate the existing fishermen for the reduction in their revenue that his fishing will bring about.[3] This free social cost of reproduction encourages overpopulation. It is bad both economically and environmentally because we would all be better off if there were fewer people consuming resources and producing pollution.

Thomas Malthus understood the tragedy of the unregulated population commons, as he showed in his spirited defense of unmarried women:

> The matron who has reared a family of ten or twelve children, and whose sons, perhaps, may be fighting the battles of their country, is apt to think that society owes her much; and this imaginary debt society is, in general, fully inclined to acknowledge. But if the subject be fairly considered, and the respected matron weighed in the scales of justice against the neglected old maid, it is possible that the matron might kick the beam. She will appear rather in the character of a monopolist than of a great benefactor to the state. If she had not married and had so many children, other members of the society might have enjoyed this satisfaction; and there is no particular reason for supposing that her sons would fight better for her country than the sons of other women. She has therefore rather subtracted from, than added to,

the happiness of the other parts of society. The old maid, on the contrary, has exalted others by depressing herself. Her self-denial has made room for another marriage, without any additional distress; and she has not, like the generality of men, in avoiding one error, fallen into its opposite.[4]

Under the two-credit system, there would be no more matrons with numerous children, because all children would have to be paid for with credits. The two-credit system would eliminate the tragedy of the unregulated commons by preventing the ultimate source of the tragedy: too many people.

In a crowded world of less than perfect human beings, mutual ruin is inevitable if there are no controls. This is the tragedy of the commons.

—Garrett Hardin[5]

Notes

1. Garrett Hardin, "The Tragedy of the Commons," *Science* 162: 1243–48, (1968).

2. John Stuart Mill, *Principles of Political Economy* (New York: Prometheus Books, 2004), 729. Originally published in 1848.

3. When the Vietnam War ended, many thousands of Vietnamese immigrated to the US. Some took up shrimp fishing near Galveston, Texas. The established American shrimpers resented this competition because it reduced their revenue. The Ku Klux Klan, led by Louis Beam, got involved and carried out operations to intimidate the Vietnamese, including burning some boats. They

ultimately failed, and by 1984, the Vietnamese dominated shrimp fishing in the Galveston area.

4. Thomas Robert Malthus, *An Essay on the Principle of Population* (1798), book 4, chapter 9.

5. Garrett Hardin, "Lifeboat Ethics: The Case Against Helping the Poor," *Psychology Today*, September 1974.

Shortage or Longage?

The ecologist Garrett Hardin pointed out that there can never be a shortage of anything without a corresponding longage. Too little of A implies too much of B. Take, for example, the current shortage of affordable housing in the United States. The corresponding longage in this case is all the people in need of affordable housing. This raises a key question: Would it be more effective to address the shortage or the longage? Should we build more affordable housing, or should we reduce the number of people seeking affordable housing? Or both? If we build more affordable housing (which we could do, for example, by changing residential zoning from single family to multifamily), we will never solve the housing shortage. That's because housing stock grows linearly, while populations grow exponentially. We would end up with even more ill-housed and homeless people. On the other hand, if we addressed the longage by reducing population size (which we could do quickly and easily by ending mass immigration and by making long-term contraception and abortion readily available), the housing shortage would soon resolve itself. The population would begin to shrink, and as it shrank, more housing would become available. That would lower housing prices and rental rates, making housing increasingly affordable for people in the lower income brackets. It would also protect farmland and natural areas from residential development, while preventing the wasteful consumption of building materials and the accompanying pollution.

But why not just raise incomes so that more people could afford housing? While it is true that raising incomes would reduce the housing shortage by stimulating housing construction, it would also stimulate reproduction, and that, in the long run, would eventually raise housing costs. In other words, we cannot escape the affordable housing shortage by either building more housing or raising incomes. The only way to permanently end it is to reverse population growth.

As a people, we have been conditioned to perceive and respond to shortages, while ignoring longages. There is a reason for this: shortages create economic demand, which produces profit for businessmen. Longages, on the other hand, are not profitable. But if our species is ever to have adequate food and water, clean air, health, peace, and a stable climate, we will have to address longages. And the place to start is with the longage behind all the shortages: the longage of people.

In the table on the following page are a few more examples of contemporary shortages and their corresponding longages. In each case, the only way to permanently eliminate the shortage is to eliminate the longage.

Shortage of Supply	Longage of Demand	Solution
Food (and sufficient fertile land to grow it)	Too many people who can't afford food; too little fertile land	Reduce birth rates and eliminate poverty. (The latter can be readily achieved once birth rates are brought under control)
Fresh water	Too many people living in arid regions and growing water-intensive crops and consuming water-intensive foods (such as beef) and wasting water	Reduce population size, relocate people to more humid regions, grow climate-appropriate crops, cut meat consumption, reduce water wastage
Well-paid jobs	Too many people seeking well-paid jobs	Reduce the labor supply by ending mass immigration. This will increase demand for labor, revive labor unions, raise wages, and enable more blacks and former prisoners to find employment.
Health care workers	Too many sick people	Provide incentives for good diet and exercise. Reduce stress, depression, addiction, and gun violence by eliminating poverty.
Free-flowing traffic	Too many vehicles	Reduce population size and establish efficient rail systems. (Road systems grow linearly; populations grow exponentially.)
Prison guards	Too many prisoners	Eliminate victimless crime laws; halt convictions of the innocent; eliminate poverty by means of a universal basic income; shorten sentences when safe to do so; issue more pardons; eliminate job and housing discrimination against former prisoners by creating a tight labor market.
Teachers	Too many students	Halt mass immigration, which is the prime source of new students in the US.

Solving shortages by reducing population size is not a new idea: More than 200 years ago, Thomas Malthus urged us to recognize that we cannot indefinitely provide for an exponentially expanding population:

> "In an endeavour to raise the quantity of provisions to the number of consumers, in any country, our attention would naturally be first directed to the increasing of the absolute quantity of provisions; but finding that as fast as we did this, the number of consumers more than kept pace with it, and that, with all our exertions, we were still as far as ever behind, we should be convinced that our efforts, directed only in this way, would never succeed. It would appear to be setting the tortoise to catch the hare.[1]

Notes

1. Malthus, Thomas. *An Essay on the Principle of Population*, edited by Donald Winch. Cambridge University Press, 1992, 229–30.

Nature's Harsh Checks

What most frequently meets our view (and occasions complaint), is our teeming population: our numbers are burdensome to the world, which can hardly supply us from its natural elements; our wants grow more and more keen, and our complaints more bitter in all mouths, whilst Nature fails in affording us her usual sustenance. In very deed, pestilence, and famine, and wars, and earthquakes have to be regarded as a remedy for nations, as the means of pruning the luxuriance of the human race.

—Tertullian, 3rd century[1]

When life is cheap, death is rich.

—Edward Abbey[2]

Pandemics of smallpox, cholera, typhus, typhoid, brucellosis, and bubonic plague once played an ugly but useful role in restoring living space in overpopulated countries. Bubonic plague, for example, killed 75 to 200 million people worldwide in the 14th century (including a large portion of Europe's population). Chronic diseases like malaria have also played an important role in helping to reduce population size.

But nowadays, major epidemics have largely been eradicated, thanks to vaccines, antibiotics, and sanitation. Even diseases that still take a heavy toll, like malaria, tuberculosis, AIDS, and now COVID-19, are less lethal than formerly, due to better drugs. As long as the world's public health systems remain intact, and new vaccines and new drugs can be developed more rapidly than viruses, bacteria, and protozoa can evolve to defeat them, mankind will remain free from truly devastating epidemics. Nevertheless, occasional outbreaks of diseases like SARS, H5N1 bird flu, Ebola, Zika, and COVID-19 remind us of our vulnerability. Especially worrisome is the growing resistance of antibiotics to harmful bacteria—a consequence of our indiscriminate use of antibiotics on livestock and unrestricted over-the-counter sales in third world countries, as well as overprescribing nearly everywhere. If we allow our antibiotics to become ineffective, even slight wounds and minor surgeries will often prove fatal. Death rates will soar.

Another grave concern is a new one: scientists can now create lethal viruses that are highly contagious. Such viruses could inadvertently escape a laboratory or be deliberately introduced by someone who has given up on mankind. The latter scenario was explored by Margaret Atwood in her novel *The Year of the Flood*.

Starvation is another reducer of populations that have grown too large. Starvation primarily kills the very young and the old. There are many proximal causes of starvation, such as droughts, floods, locust plagues, wars, plant diseases, and profiteering. But the ultimate cause is always a population that has grown too large for the long-term carrying capacity of its land.

In the 20th century, three developments greatly reduced the frequency and severity of famines. First was the discovery in 1914 of the Haber-Bosch process for converting natural gas into ammonia fertilizer. This vastly increased agricultural yields by enabling farmers to grow crops in the same fields year after year without having to fallow or use cover crops. If synthetic fertilizer were to disappear today, about 40 percent of the world's people would starve to death.[3] The second major advance against hunger was the Green Revolution of the 1960s, which introduced high-yield varieties of rice, maize, and wheat. The Green Revolution fueled a population explosion that has now made billions of people dependent on these grains and the artificial fertilizers they require. The third major development was the creation of an international system of charitable food distribution.[4]

Sadly, these dramatic increases in the food supply were not accompanied by effective birth control programs. The increase in the food supply merely encouraged further population growth, thereby setting the stage for even worse famines than those they forestalled. This was exactly what one of the pioneers of the Green Revolution, Norman Borlaug, warned against in his 1970 Nobel acceptance speech:

> The Green Revolution has won a temporary success in man's war against hunger and deprivation; it has given man a breathing space. If fully implemented, the revolution can provide sufficient food for sustenance during the next three decades. But the frightening power of human reproduction must also be curbed; otherwise the success of the Green Revolution will be ephemeral only.

Borlaug's warning echoed one issued 170 years earlier by Malthus:

> It is not in the nature of things that any perma-
> nent and general improvement in the condition of
> the poor can be effected without an increase in the
> preventive check [lower birth rates]: and, unless this
> take place, either with or without our efforts, every-
> thing that is done for the poor must be temporary
> and partial: a diminution of mortality at present will
> be balanced by an increased mortality in future; and
> the improvement of their condition in one place will
> proportionally depress it in another. This is a truth
> so important, and so little understood, that it can
> scarcely be too often insisted on.[5]

Globally, per capita food production has been slowly declining since 1985, even as population growth has soared.[6] Today, about one in seven humans is hungry and malnourished; and as the world's population continues to grow, per capita food production continues to shrink due to loss of arable land from erosion, ur-banization, salinization, and aquifer depletion. Future harvests are further jeopardized by increased resistance of pests to agricultur-al chemicals, loss of pollinators, social upheavals and wars (which disrupt distribution of food, fertilizers, and pesticides), invasive species (which are bio-pollutants that multiply exponentially), the impending exhaustion of the world's phosphorus reserves, and climate change (longer droughts, greater floods, rising sea levels, disappearance of glaciers that feed rivers used for irrigation, and higher temperatures that reduce grain yields about 10 percent for each temperature increase of 1 degree Celsius). Over the past 1,000 years, mankind has destroyed more cultivated land than it currently cultivates.[7] Clearly, the likelihood of large-scale famines in this century is high and rising.

It is noteworthy that starvation may be brought about not only by poor harvests (and losses during storage and transportation), but also by the decision of those in power to export food for profit rather than distribute it to the hungry. Ireland, for example, was a net exporter of food throughout the potato famine of the late 1840s during which 1 million starved to death and another 1 million emigrated.

More than 200 years ago, Thomas Malthus pointed out how we could permanently end famine:

> Finding therefore that, from the laws of nature, we could not proportion the food to the population, our next attempt should naturally be to proportion the population to the food.[8]

There is only one sensible and compassionate way to proportion the population to the food, and that is to practice birth control. Unfortunately, mankind has traditionally preferred another approach: warfare. Warfare increases the mortality rate and decreases the birth rate. Birth rates fall during war because many men are away from home, and the uncertainty and disruption of war discourage reproduction. Even after a war ends, destroyed infrastructure and national debt serve to depress population growth.

But for the victors, war may stimulate population growth. This is especially true when new territories and new markets are captured.

By this means the Romans built their colonies; for, feeling their city growing immoderately, they would relieve it of the least necessary people and send them to inhabit and cultivate their conquered lands. Sometimes also they deliberately fostered wars with certain of their enemies, not only to keep their men in condition, for fear that idleness, mother of corruption, might bring them some worse mischief—

> 'We bear the evils of long peace fiercer than
> war. Luxury weighs us down.' —Juvenal

—but also to serve as a bloodletting for their republic and to cool off a bit the too vehement heat of their young men, to prune and clear the branches of that too lustily proliferating stock.

—Michel de Montaigne[9]

Whenever overpopulated societies lack external enemies, they invent internal ones. Members of rival religious sects, political factions, linguistic groups, street gangs, and races turn against one another. The ensuing slaughter provides demographic relief, but only for a short while, because the underlying problem of excessive births is never dealt with.[10] On the other hand, if mankind were finally to begin regulating birth rates:

> It might fairly be expected that war, that great pest of
> the human race, would, under such circumstances,

soon cease to extend its ravages so widely, and so frequently, as it does at present.

—Thomas Malthus[11]

Human sacrifice is another practice mankind developed to ease population pressure. We tend to think of 16th century Aztecs in this regard, but mass murder on a far grander scale was carried out in the 20th century by Germany, Russia, and China. Even the Christian Church experimented with human sacrifice, hanging and burning thousands of heretics and "witches" in autos-da-fé. But nowadays, human sacrifice has fallen out of favor, and most countries have banned it. One nation that still clings to the practice is the United States, where some 2,500 people are currently awaiting execution. In contrast to the Aztecs, who killed hundreds or even thousands of prisoners in great public spectacles, the US kills its prisoners one by one, furtively, behind locked doors. Such petty killing has no significant impact on population size.

Another way population growth is checked is through sexually transmitted diseases (STDs). Gonorrhea, for example, often damages fallopian tubes, rendering women sterile. In Gabon a remarkable one-third of women are sterile due to STDs. One reason STDs are so prevalent in Gabon is that village headmen have a traditional right to copulate with any young woman they choose, and these headmen commonly carry venereal diseases.[12]

Humans are also killed by the pollution we produce. The World Health Organization reports that air pollution is responsible for about 8 million of the world's 60 million annual deaths. Most of these pollution-caused deaths are due to the burning of fossil fuels.[13]

Ecologists have gradually come to recognize that periodic fires in forests and prairies should not be suppressed, because they are necessary for ecosystem health (even though they are disastrous for the individuals who lose their habitats or are burned to death). Likewise, ecologists have come to understand that natural predators should not be exterminated, because they keep herbivore populations and their habitats healthy (even though predation is horrible for its victims). In the same way, famines, epidemics, wars, and massacres are restorative processes for human populations that have grown too large. As Tertullian recognized, mass death allows succeeding generations to enjoy a much better life—until they too overpopulate.[14]

But long before nature's harshest checks—famine, war, and epidemics—have made their appearance, subtler checks created by population pressure will already have been at work. These subtler checks include low wages, high unemployment, lack of affordable housing, food insecurity, political turmoil, high crime rates, high stress levels, low sperm counts, mental depression, and drug addiction. In former times, these stresses contributed to high rates of infant abandonment, but nowadays they persuade many women and men to use contraceptives, sterilization, and abortion, thereby lowering birth rates before famine, epidemics, and war can take hold.

The two-credit system would replace the violence of nature and human nature with a rational system to regulate birth rates and immigration. It would replace fear, pain, and hatred with reason and compassion. The ancient instinctual part of our brains would finally be harnessed by the more recently evolved prefrontal cortex. With reason more securely in charge, our descendants could enjoy a much better life far into the future. Never again would mass premature death be necessary to restore ecological and societal well-being. As Thomas Malthus recognized,

> If we be intemperate in eating and drinking our health is disordered; if we indulge the transports of anger, we seldom fail to commit acts of which we afterwards repent; if we multiply too fast, we die miserably of poverty and contagious diseases.[15]

Notes

1. Tertullian, "Further Refutation of the Pythagorean Theory: The State of Contemporary Civilisation" in *De Anima*, trans. Peter Holmes, chapter Can be viewed online at www.tertullian.org/anf/anf03/anf03-2Htm#P2978_1059891.

2. Edward Abbey, *One Life at a Time, Please* (New York: Henry Holt & Company), 128.

3. Ammonium nitrate is not only a fertilizer, but a powerful explosive. It is used in weapons of war and terrorist bombs (such as the bomb that blew up the Alfred P. Murrah Federal Building in Oklahoma City in 1995). There have also been numerous accidental explosions of stored ammonium nitrate, such as the devastating blast in Beirut in 2020. Man-made fertilizer is also the primary cause of the dead zones that are forming in coastal regions around the world. The invention of artificial fertilizer was heralded as a great benefit but turned out to be a curse.

4. Garrett Hardin observed that "world food banks move food to the people, hastening the exhaustion of the environment of the poor countries. Unrestricted immigration, on the other hand, moves people to the food, thus speeding up the destruction of

the environment of the rich countries." "Lifeboat Ethics: the Case Against Helping the Poor," *Psychology Today*, September 1974.

5. Thomas Malthus, *An Essay on the Principle of Population*, ed. Donald Winch (Cambridge University Press, 1992), 320. Twentieth-century economist Kenneth Boulding expressed the same idea this way: "If the only thing which can check the growth of population is starvation and misery, then the ultimate result of any technological improvement is to enable a larger number of people to live in misery than before and hence to increase the total sum of human misery." Boulding called this the "utterly dismal theorem." (*The Meaning of the 20th Century*, page 127).

6. Donella Meadows, Dennis Meadows, and Jørgen Randers, *Limits to Growth: The 30-Year Update* (White River Junction, VT: Chelsea Green Publishing, 2004), 57.

7. Ibid., 61.

8. Thomas Robert Malthus, *An Essay on the Principle of Population* (1798), book 4, chapter 3.

9. *Michel de Montaigne: The Complete Works*, trans. Donald M. Frame (New York: Alfred A. Knopf, 1943), 628.

10. For an informative account of the link between overpopulation and genocide, see the chapter "Malthus in Africa: Rwanda's Genocide" in Jared Diamond's book *Collapse: How Societies Choose to Fail or Succeed.* We see the same dynamic in contemporary America, where overcrowded and impoverished urban neighborhoods are experiencing civil war as young gang members shoot one another by the thousands every year.

11. Donald Winch, ed., *An Essay on the Principle of Population* (Cambridge University Press, 1992), 222.

12. *Ever Since Adam and Eve* by Malcolm Potts and Roger Short (p. 223). Susceptibility to STDs (and thus infertility) increases when the bacterial ecosystem of the vagina is upset—a condition called

bacterial vaginosis. More than one-quarter of American women have this condition at any given time. Not only does vaginosis increase the likelihood of contracting and spreading HIV and other STDs, but it also increases the likelihood of premature birth. It is treatable with the antibiotic metronidazole followed by the probiotic Lactin-V. The latter is a freeze-dried preparation of *Lactobacillus crispatus* in tampon form. To help prevent vaginosis, physicians recommend avoiding douching and using only water-based sex lubricants (never petroleum jelly). See Kendall Powell, "The Superhero in the Vagina," *The Atlantic*, October 12, 2016: https://www.theatlantic. com/health/archive/2016/10/the-superhero-in-the-vagina/503720/

13. "Air Pollution, 2020," WHO, www.who.int/health-topics/ airpollution#tab=tab_1.

14. Just as overproduction of people leads to famine, pestilence, and war, excessive debt leads to bank failures, bankruptcy, debt repudiation, and inflation. In both cases, the painful consequences restore balance, but all the pain could have been avoided in the first place.

15. Donald Winch, ed., *An Essay on the Principle of Population* (Cambridge University Press, 1992), 208.

Sobering Example of Easter Island

The confined populations of remote islands make the consequences of overpopulation obvious and hard to escape. The following account of Easter Island's ecological and social collapse is condensed from an account in Jared Diamond's book *Collapse: How Societies Choose to Fail or Succeed.* Easter Island's breakdown is a stark example of the tragedy of the unregulated commons.

Probably no human-inhabited island is more remote than Easter Island. This 63-square-mile island in the South Pacific is famous for its nearly 900 giant stone heads. It was colonized by Polynesians some 800 years ago or perhaps somewhat earlier. It is likely that only a single party of colonizers ever reached the island. At that time, Easter Island was almost entirely forested, with more than 20 species of trees. It was also one of the world's foremost seabird breeding centers, with at least 21 species. In addition to seabirds, the island had five land birds that occurred nowhere else. In the presence of such abundant resources, the Polynesian colonists and their descendants thrived and multiplied. But their success came at an extreme environmental cost. Every one of the island's tree species (including the largest palm in the world) was wiped out. All five of the endemic land birds were extirpated. No seabirds continued to nest on the island (and only five continued to nest on rocky offshore islets). The fragile volcanic soil, stripped of its tree cover, suffered severe erosion from strong winds. This erosion, combined with the loss of guano fertilizer that followed the extermination of the seabirds, led to falling agricultural output. The inhabitants also lost access to the resources of the sea because they couldn't make seaworthy canoes without trees.

Having heedlessly deprived themselves of the resources of the sea, and having destroyed the seabird populations, the islanders were forced to rely for protein on their two introduced animals: rats and chickens. The cumulative ecological stresses of overpopulation and environmental degradation finally led to prolonged civil war. The wars ended with the arrival of the Europeans, who brought lethal epidemics and slave raids. By the late 19th century, the Polynesian population of Easter Island had been reduced to about 100 people from a onetime high of perhaps 10,000.

Easter Island stands as one of the starkest reminders of what can happen when a population fails to act in time to regulate its numbers and protect the fragile resources on which it depends. Sadly, Easter Island is far from unique: with remarkable frequency, societies destroy the resources on which they depend and thereby bring about their own downfall. Jared Diamond's book *Collapse: Why Some Societies Fail and Others Succeed* contains fascinating examples of such failures as well as a few inspiring accounts of societies that were able to control their population size and rein in environmental destruction, thereby securing good futures for themselves.

James Madison, fourth president of the United States and principal author of the US Constitution, in his address to the Agricultural Society of Albemarle, Virginia, delivered on May 12, 1818, argued for an ecological perspective that would prevent the kind of calamity that the Easter Islanders brought upon themselves. Madison prophetically warned against excessive population size and excessive exploitation of resources:

> But although no determinate limit presents itself to the increase of food, and to a population commensurate with it, other than the limited productiveness of the earth itself, we can scarcely be warranted in supposing that all the productive powers of its surface can be made subservient to the use of man, in exclusion of all the plants and animals not entering into his stock of subsistence; that all the elements and combinations of elements in the earth, the atmosphere, and the water, which now support such various and such numerous descriptions of created beings, animate and inanimate, could be withdrawn from that general destination, and appropriated to the exclusive support and increase of the human part of the creation; so that the whole habitable earth should be as full of people as the spots most crowded now are...

Two-hundred years later, Herman Daly expressed the same idea from an economist's perspective:

> As the economy expands physically, it assimilates into itself an ever-greater proportion of the total life space and the total matter / energy of the ecosystem. Less is therefore available to all other species to provide the services we depend upon, such as photo-

synthesis, to mention only the most important. At some point well before the boundaries of the growing subsystem coincide with the total system, we will have sacrificed life-support services that are far more valuable than the extra commodity services that we got in return.[1]

Notes

1. Herman Daly, *Beyond Growth: The Economics of Sustainable Development* (Boston: Beacon Press, 1996), 223.

Reproductive Rights and Responsibilities

Everyone has a right to live. We will suppose this granted. But no one has a right to bring creatures into life, to be supported by other people.

—John Stuart Mill[1]

Advocates for an unlimited right to breed sometimes cite the Proclamation of Teheran, a 1968 United Nations human rights document that stated that "parents have a basic human right to determine freely and responsibly the number and spacing of their children."[2] The key word here is "responsibly"—a word that modifies "freely." To produce more than two children in an overpopulated world without first legitimately acquiring the reproductive rights of someone else is inherently irresponsible. Therefore, it is wrong to infer from the Proclamation of Teheran that people have an unlimited right to breed. In fact, overbreeding is the single greatest threat to human rights, because it leads to poverty, famine, pestilence, war, genocide, and the denial of a good life to future generations. Moreover, when people in an overpopulated country produce more than two children, they deprive some of their fellow citizens and their very own descendants of the opportunity to have children. John Stuart Mill explained this with mathematical precision:

> In a country where population has no room to increase, or in which its progress must be so slow as to be hardly perceptible, when there are no places vacant for new establishments, a father who has eight children must expect, either that six of them will die in childhood, or that three men and three women among his contemporaries, and in the next generation three of his sons and three of his daughters, will remain unmarried on his account.[3]

The two-credit system is in complete accord with the Proclamation of Teheran, because it allows parents to "determine freely and responsibly the number and spacing of their children." *Responsibly* means that if parents want more than two children, they would first have to acquire additional credits from their fellow citizens. The two-credit system actually goes one better than the Proclamation of Teheran, because it would also protect the reproductive rights of nonhuman animals and plants, by preventing their habitats from being overrun by an exponentially expanding human population.[4]

Curiously, advocates for an unlimited right to reproduce rarely assert that their nation must accept an unlimited number of immigrants. Somehow, they are able to discern the dangers of unregulated immigration (and are willing to support coercive measures to restrict it) while remaining blind to the equivalent dangers of unregulated reproduction.

To those who assert that people have a God-given right to both unlimited reproduction and food, Garrett Hardin responded that this would imply that God "must be bent on the utter destruction of civilization."[5]

Economist Herman Daly recognized that what we really need is not a human right to breed without limits, but a human right to be born into a society that is not already overpopulated.[6]

Notes

1. John Stuart Mill, *Principles of Political Economy* (New York: Prometheus Books, 2004), 352.

2. "Proclamation of Teheran," International Conference on Human Rights, 1968.

3. John Stuart Mill, *Principles of Political Economy* (New York: Prometheus Books, 2004), 362.

4. It is impossible for people to "determine freely and responsibly the number and spacing of their children" if they are ignorant of the basic facts of demography, economics, and contraception. Informed decisions require informed minds. Malthus drew attention to this: "And till this obscurity is entirely removed, and the poor are undeceived with respect to the principal cause of their past poverty, and taught to know that their future happiness or misery must depend chiefly upon themselves, it cannot be said that, with regard to the great question of marriage or celibacy, we leave every man to his own free and fair choice." (Donald Winch, ed., *An Essay on the Principle of Population*, Cambridge University Press, 1992, 242.)

5. Garrett Hardin, *Naked Emperors: Essays of a Taboo-Stalker* (Los Altos, CA: William Kaufmann, Inc., 1982), 261.

6. Herman Daly, *Beyond Growth: The Economics of Sustainable Development* (Boston: Beacon Press, 1996), 120.

Why a Voluntary Approach to Population Reduction Can Never Succeed in the Long Run

Advocates for a voluntary approach to population reduction point out that fertility rates in many countries have already fallen below the replacement level of 2.1 children per woman. These countries include Brazil, Russia, Armenia, Chile, Cuba, Iran, Australia, Canada, United States, Japan, North Korea, South Korea, Vietnam, and the nations of Europe.[1] However, many of these countries (including the US) are still growing rapidly due to population momentum and/or immigration. Halting population growth would require lowering birth rates even further and/or spacing babies farther apart and restricting immigration. But even if a nation truly halted population growth (as Japan has done), a voluntary reduction in birth rates could never persist for long once the population fell below its carrying capacity. Here are the reasons why:

1. People who use contraception produce fewer children than those who do not. As a result, their genes diminish in frequency in the gene pool, while the genes of those who do not use contraception come to prevail. Any genes that strengthen the sex drive, strengthen the longing for babies, or strengthen the pride and pleasure that people

derive from having children, will be favored by natural selection.[2] Natural selection will also favor cultural values that encourage early marriage and large families. Therefore, in the absence of effective regulation, populations will continue to grow until they exceed their carrying capacity, whereupon stress, hunger, disease, and war will check further growth.

2. In many countries that currently have relatively low birth rates (including the US), the main factor keeping people from having more children is cost. During economic recessions and depressions, birth rates always drop, and during good times they always rise. Polls consistently indicate that most Americans would like to have an average of 2.6 children (0.5 above the replacement level).[3] In other words, Americans are not "volunteering" to have small families; they would prefer larger families. During the period 1946–64, when most Americans were economically well off, they produced the post-war baby boom. And during the 17th and 18th centuries, when Americans enjoyed a seemingly endless supply of cheap, fertile land, they produced the largest baby boom in history. The prevailing desire of Americans and other peoples for above-replacement-level fertility exposes a critical weakness in the idea that a voluntary approach to population reduction could work in the long run. Malthus understood this, declaring that if we supposed "no anxiety about the future support of children to exist, I do not conceive that there would be one woman in a hundred of twenty-three years of age, without a family."[4]

3. The relative ease with which the sexual and maternal in-
    stincts overwhelm sound judgment is demonstrated by
    the fact that 18 percent of US pregnancies are complete-
    ly unwanted and 27 percent are mistimed. Add them
    together and 45 percent—nearly half—of all US preg-
    nancies are unplanned.[5] This is true despite the fact that
    most American women are educated and have access to
    effective contraception. Fifty-eight percent of these un-
    planned pregnancies end in births (excluding miscar-
    riages and stillbirths) and 42 percent end in abortions.[6]
    That means that 26 percent of all US *births* are the result
    of unintended conceptions ($0.58 \times 0.45 = 0.26$). This
    undermines the argument that a voluntary approach to
    population reduction could be successful. For too many
    people, the instinct to reproduce is so strong that only
    long-term contraception and social regulation can effec-
    tively constrain it.

4. In nations like Iran and Russia, which have succeeded
    in greatly lowering their birth rates by voluntary means,
    we now see their governments trying to increase birth
    rates. The Iranian government has reduced its funding
    for birth control programs, while the Russian govern-
    ment offers cash bonuses of $10,000 for each baby. Just
    because birth rates drop doesn't mean they will continue
    to drop. As long as there is a niche that humans can fill,
    they will fill it with more babies (or more immigrants) if
    not prevented by an effective system of population con-
    trol.

5. The fertility rate in the United States would be substantially higher if hundreds of thousands of young, fertile Americans were not locked up in prisons during their prime reproductive years. These prisoners are not "voluntarily" restricting their fertility. In the US, dark-skinned people are disproportionately represented in prison populations. Mass imprisonment therefore functions as a race-based fertility reduction program.

6. There are many countries where a large proportion of the population regards contraception as immoral. In Pakistan, for example, 65 percent of the population disapproves of contraception, while in Nigeria the figure is 54 percent.[7] In such cultures, honor and respect are conferred upon those who produce the most children (in part because children are valued for their labor and as future support for parents). A voluntary approach to population reduction is doomed to fail when a large share of the population regards contraception as immoral and regards large families as desirable.

7. When a population begins to shrink, the labor supply tightens. This creates pressure to raise wages. But employers don't want to pay higher wages, so they prevail upon their government to admit more immigrants. The arrival of these immigrants quickly cancels out the incipient population reduction and keeps wages low, especially when immigrants have higher birth rates than natives. We see this happening today in the United States, Great Britain, Germany, and Sweden.

8. When an economy collapses, contraception and abortion may become unaffordable, causing birth rates to rise. This pushes families into even deeper poverty. That is the situation today in Venezuela, where a condom costs more than a typical day's wages. This is another reason why the copper IUD is the best form of contraception.[8]

For all the above reasons—but especially the first one—a voluntary approach to population reduction can never succeed in the long run. Nature would never allow it. That is why birth rates and immigration must be rationally regulated.

Garrett Hardin pointed out that democratic societies are based on the principle of *mutual coercion mutually agreed upon*, and that to reject such coercion is to reject democracy itself.[9] The alternative to population regulation is anarchy, which is what America has today with respect to reproduction and, to a large extent, immigration. Two examples of the latter are the failure of the US government to halt illegal immigration and its failure to impose a quota on the largest category of legal immigrants: family members brought to the US under the "family reunification" program.

Until recently, many demographers believed that the key to lowering birth rates was for poor countries to become industrialized. They based this idea on the observation that birth rates fell in Europe and the United States following industrialization. They assumed that the same "demographic transition" would take place in third world countries. This notion of a natural demographic transition was welcomed by the world's leaders because they assumed it meant they could sit back and let the population problem solve itself. One country that had the good sense not to sit on its hands was China. Instead of waiting for industrialization to lower its birth rate, China first lowered its birth rate. That freed up enough capital to enable China to industrialize.

In the 1990s, as the limitations of the demographic-transition theory were becoming apparent, another feel-good notion took hold. This was the idea that the key to reducing birth rates was to educate women. Influential feminists embraced this idea and soon made it the "politically correct" response to overpopulation. The argument went like this: if women were educated, they would enter the workforce, and that would deprive them of the time and energy to care for a large family. But in truth, it is not education that reduces family size, but the availability and affordability of contraception and abortion. Even when there are no jobs available, most women will still choose to have small families if long-term birth control and abortion are accessible and affordable. After all, having a small family gives women more time and resources, which improves their lives as well as the lives of their children.

> Where family planning is easy to get, as in Thailand, there is little difference in family size between rich and poor, illiterate and well-educated...Here is what many people have failed to understand about family planning: Rather than reduced family size being a beneficial side effect of development, access to family planning can help drive the development.[10]

But even if all women had easy access to contraception and abortion, birth rates would still not fall fast enough to avert severe social and environmental disaster. What is also needed is wise regulation, both for rapid population reduction and for fairness. The two-credit system provides a fair balance of freedom and regulation. It is voluntary insofar as it leaves people free to determine what to do with their credits. It is regulatory insofar as it responds appropriately and effectively to violations. It is a system that can free us forever from famine, pestilence, war, and massacres and allow us to intelligently guide our future evolution so that we can become a more rational and enlightened species.

Notes

1. The CIA World Factbook (see cia.gov) lists total fertility rates by country in descending order.

2. It was Charles Darwin who first called attention to the genetic implications of the widespread use of contraception: "The advancement of the welfare of mankind is a most intricate problem: all ought to refrain from marriage who cannot avoid abject poverty for their children; for poverty is not only a great evil, but tends to its own increase by leading to recklessness in marriage. On the other hand, as Mr. Galton has remarked, if the prudent avoid marriage, whilst the reckless marry, the inferior members tend to supplant the better members of society." (*The Descent of Man*, chapter 21.)

3. Donald Winch, ed., *An Essay of the Principle of Population* (Cambridge University Press, 1992), 59.

4. See www.gallup.com/poll/164618/desire-children-norm.aspx. Despite the desire of Americans for an average of 2.6 children, the 2021 fertility rate in the US was 1.84 children per woman (www.cia.gov/the-world-factbook/field/total-fertility-rate/). It would be interesting to know how many of those who expressed a desire for more than two children did not yet have children. Presumably, the more children people have, the less they desire more children.

5. In 2011, 45 percent of the 6.1 million pregnancies in the United States were unintended. See www.guttmacher.org.

6. See www.guttmacher.org.

7. The list of countries where polls indicate that a large proportion of the population regards contraception as immoral includes Pakistan (65 percent), Nigeria (54 percent), Ghana (52 percent), Malaysia (40 percent), Tunisia (39 percent), Uganda (38 percent), Kenya (33 percent), Philippines (29 percent), Senegal (28 percent), El Salvador (23 percent), India (22 percent), and Turkey (21 percent). Source: www.pewresearch.org. It is noteworthy that all the above nations have strongly patriarchal cultures.

8. See www.nytimes.com/2021/02/20/world/americas/venezuela-birth-control-women.html.

9. Garrett Hardin, *The Ostrich Factor* (New York: Oxford University Press, 1998), 78.

10. Malcolm Potts and Thomas Hayden, *Sex and War* (Dallas: BenBella Books, 2008), 315.

Implications of Major Life Span Extension

Looking to the future, we should consider the possibility that biologists may succeed in manipulating human genes to extend longevity. In the laboratory, researchers have already extended the life span of one nematode species (*Caenorhabditis elegans*) by a factor of six.[1] By manipulating a few genes, these worms have been kept young and vigorous until their last days. Doing the same for humans would enable us to have healthy life spans of 500 to 600 years. Such long lives would enable us to reduce the amount of time and money spent educating the young and would permit those already educated to contribute much more. Having a longer lifetime might even make people wiser (one can hope). In any case, to prevent another population explosion, longer lifetimes would require that women not only keep their fertility no higher than the replacement level (two babies per woman), but also delay reproduction until much later in life, and space their children much farther apart, as in the following example:

Average Lifespan (years)	Age at First Childbirth	Spacing of Children (years)
60	25	5
600	250	50

One wonders how brothers and sisters born 50, 100, or 200 years apart would relate to one another.[2] Mankind would do well to think carefully about such issues before funding research that could lead to radical life extension. Once the genie of longevity is released, it will be extremely difficult to control. Who is going to tell a person who could live 600 years that he must die at 90 to make space for a baby?[3]

> Know when to stop, avoiding peril. That is how to live long.
>
> —*Tao Te Ching*[4]

Notes

1. For more on *Caenorhabditis elegans* longevity research, see calicolabs. com, frontiersin.org, and www.ncbi.nlm.nih.gov/pmc/articles/ PMC3001308/.

2. The authors of the biblical book of Genesis seem to have understood that if lives were extraordinarily long, people would have to postpone reproduction for many decades. Adam, whose life span was 930 years, did not beget Seth until he was 130. Seth, who lived 912 years, did not beget Enosh until he was 105. And Methuselah, who lived 969 years, did not beget Lamech until he was 187.

3. In 2015, Google set up a subsidiary called "Calico" whose task is to find ways to extend human life span. This is the kind of irresponsible research that prompted Edward Abbey to recommend

civilian review boards to prevent "the misapplications of scientific discovery" (*Postcards from Ed*, p. 136). If Google seriously wants to do something socially beneficial, let it work to reduce population size. It could do so, for example, by providing reproductive education and free IUDs and abortion to the world's poor.

4. Lao Tzu. *Tao Te Ching: A New Translation*, trans. Sam Hamill (Boston: Shambhala Publications, Inc., 2005), chapter 44.

Harnessing Natural Selection: A Revolution in Evolution

Mankind, along with all other species of life, is currently trapped in a never-ending round of population increase followed by painful population reduction. Nature first urges us to enjoy the pleasures of sex and children and then makes us suffer for the resulting overpopulation. But nature gave humans—uniquely—the capacity to rise above its harsh program. It gave us a brain than can understand the dangers of overpopulation and develop techniques to prevent or eliminate it. If we use this precious gift to reduce our population to an optimal size, we will become the first species to successfully harness natural selection, the first to escape nature's cruel round. Instead of being pushed along blindly by the forces of natural selection, we will finally be able to consciously guide our evolution.

The importance of harnessing natural selection cannot be exaggerated. For 3.6 billion years, life has been inching toward this possibility. We are the first generation of beings to have this astonishing power within our grasp. And this is happening at the very time when we are learning how to modify our genome, opening additional possibilities for rapid evolutionary advance.

> Indeed the idea that man might create his successor
> is no longer so fanciful as it seemed even twenty-five
> years ago.
>
> —Kenneth Boulding (in 1964)[1]

If we do succeed in reducing our population to an optimal size, the more extreme expressions of the "struggle for existence" will cease. There will be no more famine, pestilence, war, and massacres. Natural selection, instead of operating primarily through the mechanism of premature death, will operate mainly through sexual selection, supplemented by deliberate gene manipulation.

Sexual selection takes two quite different forms. The more familiar form occurs when two males challenge each other for sexual access to females. Such contests are characteristic of many species, including horses, elk, and elephant seals. These male-on-male competitions also occur at the cellular level when a woman has unprotected sex with two or more men in short succession. Inside her body, each man's sperm competes to be first to reach the goal.[2]

The second kind of sexual selection was described by Darwin this way:

> There is another and more peaceful kind of contest,
> in which the males endeavor to excite or allure the
> females by various charms.[3]

In this form of sexual selection, females freely choose their mates instead of being claimed by victorious males. Because females choose the more attractive, charming, and healthy males, this produces an attractive, charming, and healthy population. Again, Darwin:

The females are most excited by, or prefer pairing with, the more ornamented males, or those which are the best songsters, or play the best antics; but it is obviously probable, as has been actually observed in some cases, that they would at the same time prefer the more vigorous and lively males…In the converse and much rarer case of the males selecting particular females, it is plain that those which were the most vigorous and had conquered others, would have the freest choice; and it is almost certain that they would select vigorous as well as attractive females.[4]

In today's overpopulated, highly competitive societies, women seek men with high incomes and high social status. But in an un-crowded, egalitarian society where poverty and insecurity no longer existed, women would favor men who were good-looking, healthy, attentive, charming, responsible, fond of children, protective, and skilled at courtship and lovemaking. Under a two-credit system, this would result in the more attractive males siring two (or occasionally more) children, while the less attractive would sire one or none. By relieving men of the pressure to accumulate wealth in order to attract women, sexual selection would greatly reduce resource consumption and pollution.[5] In China's Yunnan province, the Mosuo people have already developed a culture in which sexual selection by women pre-dominates. They have shown that such a system works well.[6]

There is no goal more important than bringing the sexual and maternal drives under rational control.

Civilization in every one of its aspects is a struggle against the animal instincts. Over some even of the strongest of them, it has shown itself capable of acquiring abundant control…If it has not brought the instinct of population under as much restraint as is needful, we must remember that it has never seriously tried. What efforts it has made, have mostly been in the contrary direction.

—John Stuart Mill[7]

The two-credit system would allow us to indulge our "instinct of population" without harmful consequences to ourselves, our society, or our environment. The system would function like a sober driver, delivering everyone home safely after a night of indulgence. It would enable us, with the help of genetic engineering, to rapidly become a much-improved species, or perhaps even a new species to replace *Homo sapiens*. I hope so.

This much I think I can, and do, assert:
That our perverse vestigial native ways
Are small enough for reason to dispel.
So that it lies within our power to live
lives worthy of the gods.

—Lucretius[8]

Notes

1. Kenneth Boulding, *The Meaning of the 20th Century: The Great Transition* (New York: Harper Colophon Books, 1964), 153.

2. There is apparently good evidence that sperm compete as a team. See www.bbc.com/news/health-27589816.

3. Charles Darwin, *The Descent of Man*, chapter 18 summary.

4. Ibid., chapter 8.

5. In the future, although we will have the capacity to genetically engineer our children, most people will probably choose the same traits in their engineered offspring that they would choose by sexual selection in their mates. One trait they are unlikely to choose is a proclivity for violence.

6. Among animals, strong sexual dimorphism correlates with (1) polygyny (females fertilized by relatively few males), (2) high levels of male-on-male aggression (as they compete for females), (3) males not participating in childcare, and (4) females subjected to male control. In contrast, sexual monomorphism correlates with monogyny (or serial monogyny), peaceful relationships, males participating in childcare, and sexual parity. Humans are a weakly dimorphic species. It is our culture that determines whether we behave more like a dimorphic or monomorphic species. See Robert Sapolsky, *Behave: The Biology of Humans at Our Best and Worst*, pp. 356–57.

7. John Stuart Mill, *Principles of Political Economy* (New York: Prometheus Books, 2004), 360. Originally published in 1848.

8. Lucretius, *The Way Things Are: The De Rerum Natura of Titus Lucretius Carus*, trans. Rolfe Humphries (Bloomington: Indiana University Press, 1968), 95.

Is It Ethical to Have Children?

> Children are smarter than any of us. Know how I know that? I don't know one child with a full-time job and children.
>
> —Bill Hicks[1]

Children are extremely costly long-term parasites. Yet many of us desire children, love them, and gladly spend our time and money on them. We do so because natural selection has given us compensatory pleasures for the hardships of parenthood. Among these pleasures are the excitement and elation we feel when we are in love and the intense pleasure of the sex act. Nature intended these pleasures to lead to pregnancy—even when reason and prudence would argue against it.[2]

Another inducement to having children is the pleasure that comes from fulfilling maternal instincts. Nursing, for example, is intensely pleasurable for most mothers. The female attraction to babies is surely as strong as the male attraction to women.

Another compensatory pleasure for parenthood is the pride parents feel when they produce children. Procreation makes men feel more masculine and women more feminine, and it wins them the approval of their parents and peers. Parents also take pride in their children's achievements.

Another inducement to having children is their openness and playfulness, which make them fun to be around:

> Only a fool envies the joy of a child; a grown-up man
> or woman shares in that joy.
>
> —Edward Abbey[3]

People may also have children because they wish to experience the pleasure of giving their children the love and advantages they may never have known, or which they would like to relive vicariously.

People also become parents for a variety of dubious reasons. A woman may hope that her pregnancy will induce her boyfriend to marry her. Or she may think her pregnancy will help patch up a fraying marriage. Or she may feel lonely and want a child for companionship and love. Or she may fear she'll one day regret not having a child. (After all, so many people have told her she will.) Or she may think her first two children need siblings for companionship (or perhaps a sibling of a different sex). Or she may have reached a point in her life where she doesn't know what to do next and turns to childbearing to give her life purpose and meaning.

Besides the pleasures of parenthood, there is a practical reason to have children: all of us, if we live long enough, will become infirm and need someone to care for us. Our children are the logical caregivers, because no one else young enough to be up to the task is likely to love us enough to undertake it. When our children become our caregivers, the roles of host and parasite are reversed.[4]

Because there are both compensatory pleasures and a practical reason for parenthood, it is only natural that most people desire to have children. But what about overpopulation? Can it ever be ethical to reproduce when we are already living unsustainably? I believe it can be. In the following chapters, readers will see that the two-credit system would allow reproduction without causing population growth. In fact, the population would gradually decrease to an optimal size even as babies and immigrants were added. Moreover, the two-credit system would enable mankind to finally harness natural selection, enabling us to eliminate evil through genetic engineering (evil being the brain switch that shuts off empathy.)

Notes

1. Performance by Bill Hicks at the Laff Stop in Austin, Texas, 1991. It can be viewed at www.theguardian.com/stage/video/2014/feb/25/bill-hicks-standup-childbirth-exclusive-video.

2. Brain researchers have made considerable progress in understanding how the sex drive suppresses the brain's rational powers. They have demonstrated experimentally that when people are shown a photograph of a lover, a region of the brain called the ventral tegmental area (part of the pleasure and reward circuit) is strongly activated, while simultaneously the dorsolateral prefrontal cortex (the reasoning center) is deactivated. See *BBC News*, April 26, 2016.

The intense pleasures that lead to pregnancy are candidly described in this verse:

Her Evening Plan

We'll hug and kiss
And whisper and giggle
We'll shed our clothes
And wrestle and wriggle
We'll fondle and feel
And reveal what is real
And when I get all slick and sopping

Your sweet tongue will do the mopping
And then you'll lay me over your knee
And spank my ass to ecstasy
You'll whacks poetic like a bard
Till I'm agasp and you're so hard
And then you'll mount me (loving master!)
And pump my cunt-hole ever faster
Till neither one can hang on more
And shaking, quaking, we merge our core
And from our union
Life will come
Oh God, oh Jesus, oh what fun!

3. Edward Abbey, *One Life at a Time Please* (New York: Henry Holt & Company), 63.

4. There is an amusing Chinese tale of an old man whose son grew weary of supporting him. The son put the old man in a coffin and was about to push it off a cliff when the father knocked. The son opened the coffin, and the father said, "Son, I don't mind if you push me off the cliff, but please save this fine coffin. Your son will need it for you."

Part 2:

The Two-Credit System

Population Credits

To achieve population control (which is a need of the community) we must devise a community-mandated control system that confers tangible rewards to families and individuals who are asked to curtail their fertility for the "good of the species" (or of a large group like the nation).

—Garrett Hardin[1]

The central concept of the population reduction plan presented in this book is that every time an individual is added to the population, whether by birth or immigration, the people responsible for the addition (the parents or the immigrant) must pay two "population credits."

Citizens would be issued two population credits at birth as their birthright. Individuals could receive additional population credits as donations from friends or relatives, or through purchase from those who have decided to sell their credits. There would also be a one-time issuance of credits to those under the age of 50 at the time the plan goes into effect.

Since individuals would have two population credits as their birthright, a reproductive couple would have a total of four credits at their disposal. That would allow them to produce two children (paying two credits for each child). By producing two children, the parents would, in effect, be making future replacements for themselves. Demographers call this level of fertility the replacement level. People who prefer to have only one child, or none at all, would be free to donate their remaining credit(s) to whomever they wished, at home or abroad, or sell their credit(s) either privately or in a government-regulated online market. Anyone in the world with internet access would have the opportunity to purchase credits (though the number of credits available to foreigners would be limited to ensure that natives can continue to have access to extra credits). Any unused credits would automatically expire when people turned 50. The unused credits of anyone who dies between the ages of 18 and 50 would also expire (except in the special case explained in the chapter "Canceling the Impact of Unaccredited Births").

People who wish to have more than two children would have five options: (1) receive additional credits as a donation (most likely from a childless sibling), (2) purchase additional credits in the credit market, (3) adopt a child whose birth parents already paid the credits, (4) adopt an embryo whose sperm and egg donors paid the credits, (5) find a new mate, and use that mate's two credits to pay for an additional baby.

Just as people would need to pay two credits whenever they added a baby to the population, people would also need to sell or donate two credits to a would-be immigrant before he or she could immigrate to this country. Those who wish to immigrate would have their own electronic credit account set up at the nearest US consulate, into which their donated or purchased credits would be deposited. When the immigrant arrived at a US port of entry, citizenship would be conferred immediately, just as it is on a baby at birth. (After, of course, a check to screen out criminals and potential terrorists.) The new immigrant would be the population equivalent of a new baby.

To sum up, citizens could use their two credits in the following ways:

1. They could pool their two credits with those of a spouse/partner in order to produce two children.
2. They could donate one or both of their credits, thereby enabling the recipient to have more than two children. These donations would be officially registered and would be irrevocable, although recipients could voluntarily return the credits to the original holder.
3. They could donate one or both of their credits to a prospective immigrant. The immigrant would set up an account through the nearest US consulate and the donated credits would be deposited there.
4. They could sell one or both of their credits either privately or in a global online market operated by the government. The credits could be purchased by their compatriots who wish to have more than two children or by foreigners who wish to immigrate to this country.
5. They could choose to do nothing with their credits, letting them expire on their 50th birthday. The credits of anyone who dies between 18 and 50 would also expire. This attrition of credits would be one of the mechanisms whereby population size is reduced.

The two-credit system is analogous to nature's own binary reproductive system, which uses the union of two gametes (a sperm and an ovum) to add an individual to the population. A pair of credits accomplishes the same thing. But whereas people have numerous gametes, they would have only two population credits, allowing them to just replace themselves in the population (unless they acquired additional credits). Population credits share with gametes some other features: they can be donated, sold, or allowed to expire. Because they can be donated and sold, they can be used to regulate immigration. Population credits are a kind of currency that can be converted into babies, immigrants, or cash. And, like currency, population credits can be retired or returned to circulation.

Once in operation, this credit-based system would gradually reduce any population to a democratically determined optimal size and then keep it there. The plan permits precise adjustments to population size. For example, if an epidemic were to suddenly decimate the population, the government could put retired credits back into circulation by means of a lottery. That would enable the population to rebuild to its former size.

For a plan like this to be fair, people should never feel compelled by poverty to sell their population credits in order to obtain life's necessities. Instead, poverty itself should be eliminated. I show how this can easily be accomplished in Part 5.

The two-credit system is a kind of pollution permit system. In fact, it is by far the most effective pollution permit system because it curtails the ultimate source of pollution: people. The two-credit system would grant every citizen-polluter the right to add one new polluter (either a baby or an immigrant) to the population. The new polluter would eventually replace the one who added him or her. This is replacement-level pollution. Economist Herman Daly pointed out that for a system of pollution permits to be effective, two requirements must be met:

1. The number of rights to pollute must be limited to no more than the capacity of the ecosystem to absorb and process the pollution.
2. The rights to pollute must be distributed to the polluters in a fair manner.[2]

The two-credit system fulfills both requirements: it keeps pollution within the sustainable limits of the ecosystem by reducing the population to an optimal size and limiting the number of children to two (unless the parents legitimately acquire additional population credits); and it distributes the rights to pollute (i.e., add children or immigrants) in a fair manner, by making population credits available to all citizens as their birthright.

By keeping reproductive rights within the limits of the earth's capacity to process our pollution, and by distributing the reproductive rights equitably, the two-credit system would allow the free market to efficiently reallocate pollution rights.

Some may worry that certain religious groups would try to maximize their reproduction by insisting that their members donate any unused credits to other members of the faith; or they might prohibit members from selling their credits or allowing them to expire. If any religion chooses to do this, it would be free to do so. But in practice it may prove difficult to dissuade members from selling credits when they can reap a large profit from doing so. Moreover, only *individuals* would be able to buy credits, and there would be a 10-credit maximum on purchases. That means no individual could donate more than 12 credits (their own two birthright credits plus 10 purchased credits). This means neither rich individuals nor religious organizations would be able to purchase a large block of credits and then distribute them to those of their own persuasion. In any case, most people who already have two children are not eager to have more. If given an extra pair of credits, most people would sell them.

Thoughtful readers may ask, "Why not give credits to women only, leaving out men?" That could certainly be done, since a credit-based system to regulate birth rates and immigration would work perfectly well if only women were issued credits. A woman would still have two credits, and she would pay one credit for each baby she produced. She could also donate or sell unused credits. A female-only credit system would also be simpler and less expensive to administer than one that included men. However, a female-only system would have some serious drawbacks:

- Men who have no children, or only one child, would be deprived of the opportunity to make a substantial sum of money by selling their remaining credit(s).
- Men (including gay couples) would not be able to adopt children unless they purchased credits or received them as donations.
- The entire burden of regulating fertility would be placed on women even though men are 50 percent responsible for pregnancies. (In paternalistic societies, men are more than 50 percent responsible for pregnancies, because they pressure women to produce children and deny them contraception and abortion.)

- It would free men from all restraints on how many children they could sire. As a result, men with higher-than-average testosterone levels would produce more children. This would increase the average testosterone level in the population. Higher testosterone levels would likely result in less attentive husbands and fathers, and more societal and domestic violence—behaviors that work against the advancement of civilization.

If we want a society in which both genders have equal rights under the law, then men as well as women must be issued population credits.

Finally, it is interesting to compare the two-credit system to the solution to overpopulation proposed by Thomas Malthus. He believed that the best way to prevent overpopulation was for couples to exercise "moral restraint." By this he meant practicing temporary or permanent celibacy. He felt that moral restraint was a much better approach than contraception, abortion, infanticide, masturbation, recourse to prostitutes, anal intercourse, or allowing nature to solve the problem through famine, pestilence, and war. He opposed contraception because he feared men would become lazy if they did not have large families to support. What Malthusian moral restraint shares with the two-credit system is the goal of reducing the birth rate by prolonging celibacy. But whereas celibacy for Malthus meant keeping the genitals of males and females apart (chastity), celibacy under the two-credit system means keeping sperm and egg cells apart by means of modern contraceptives. In other words, the two-credit system represents moral restraint through gametic chastity. By eliminating the requirement for prolonged sexual abstinence, the two-credit system provides a solution to overpopulation that is much more palatable, and far more effective, than the voluntary chastity urged by Malthus and the Catholic Church.

Notes

1. Garrett Hardin, *Living within Limits* (New York: Oxford University Press, 1993), 239–40.

2. Herman Daly, *Beyond Growth: The Economics of Sustainable Development* (Boston: Beacon Press, 1996), 52–53.

Three Ways to Pay Two Credits

Whenever a baby is born, the parents could pay the required two credits in one of three ways: (1) the father could pay both credits from his account; (2) the mother could pay both credits from her account; or (3) each parent could pay one credit from their respective accounts. Parents would decide for themselves how to divide the credit payment. If the parents can't agree, then one credit would automatically be deducted from each parent's account. If one of the parents had no credits, then both credits would automatically be deducted from the other parent's account. The number of credits paid by each parent would be recorded in that parent's electronic credit account. If the parents do not have two credits to pay for the child, the procedures described in the chapter "Curbing Unaccredited Births" would be followed.

One problem that could arise in the payment of credits is that a woman who had responsibly checked to make sure that she and her mate had enough credits to pay for a baby might discover upon giving birth that at some point during her pregnancy her mate secretly sold his credit(s) or used them to pay for a baby he conceived with another woman. To prevent this misfortune, all women would have the ability to put an electronic lock on their mate's credits before engaging in unprotected sex with him. Her mate would, of course, have to agree to this lock, and if he refused, the woman would know immediately that he was untrustworthy. This locking of credits could be accomplished through a secure online process. Placing a lock on credits is akin to placing a restraining order on assets by a divorce judge. Divorce judges do this to prevent dishonest spouses from divesting or hiding their assets.

Inviolability of Population Credits

Population credits would be an inalienable birthright. The government would not be allowed to deprive anyone of their credits for any reason. Even prisoners under life sentence would retain their credits and be free to donate them or sell them. The government could not compel prisoners to sell their credits to pay for the prisoner's keep, to pay a fine, or to pay restitution to victims of a crime. By retaining control over the proceeds of their credit sales, prisoners could, if they wished, provide income for a spouse or children; and, once released, they could use the money to help support themselves—an important consideration because many prisoners lack job skills and face strong job discrimination. This would lower the crime rate.

Another reason the government should be prohibited from compelling prisoners to sell their credits is that this would constitute a major violation of human rights, comparable to forced sterilization. Most prisoners belong to racial and ethnic minorities, so depriving them of their credits would constitute a form of "ethnic cleansing." In fact, imprisonment itself restricts reproduction by ethnic minorities.

Just as the government would not be allowed to appropriate anyone's credit proceeds, neither could it withhold credits from the mentally ill or physically defective—whether babies or immigrants. The ugly history of eugenics in the United States and Germany provides ample evidence that governments should not be trusted with schemes to "improve the race."

The proceeds from the sale of credits would be exempt from all federal taxes. There are two reasons for this. First, population credits would be every citizen's birthright, so the government should have no claim on any part of their economic value. Second, politicians shouldn't be allowed to use tax policies to manipulate how citizens use their credits. For example, if a heavy tax were imposed on the proceeds from the sale of population credits, citizens would probably be less inclined to sell credits. Instead, they might decide to use them for reproduction. This would mean more domestic births. Although politicians, for whatever reason, might think this was a good idea, it is fundamentally none of their business. Such decisions should be made exclusively by the individuals who hold the credits. The population credit market should be a free market.

Registration Process

Registration would be required for five groups: (1) newborn babies, (2) adults who plan to have children and therefore need credits, (3) those who would like to obtain credits in order to donate or sell them, (4) prospective immigrants who would like to purchase credits or receive them as donations, and (5) resident illegal immigrants and resident asylum seekers (both of whom would become legal US citizens upon registering but would not receive population credits). Eventually, the entire population would be registered, because all babies would be registered at birth.

At least one trained and certified registrar would be attached to every hospital that has a maternity ward. In most cases, this official would be a physician already affiliated with the hospital who had received registrar training. Besides registering newborns, the registrar would be responsible for registering adults, including resident illegal immigrants. People in other countries who wish to immigrate would be registered at the nearest consulate by an authorized consular official.

At the time the plan goes into effect, adults under the age of 50 who have not produced any children would have two credits deposited in their accounts upon registration. Adults under 50 who have produced one child would receive one credit. Adults under 50 who have produced two or more children would receive no credits.

Resident illegal immigrants would receive no population credits, but they would be granted the opportunity to become citizens. If they decline this opportunity they would be deported.

The registration process would consist of two simple steps: (1) provide proof of identity and (2) provide a saliva sample for DNA record keeping. The purpose of the DNA sample is to establish a genetic database that would be used to determine the true parentage of every newborn baby. Citizens could prove their identity by means of a passport, secure driver's license, or other reliable means. In the case of people who do not already possess a secure form of identification, the government would determine the validity of their self-declared identity and provide them with a passport or identification card. This service would be free.

The fact that population credits would be worth a great deal of money would inevitably tempt some parents to falsely claim they had not produced children in order to obtain credits. To deter such fraud, the following measures would be taken:

- Applicants would be required to sign a statement declaring that their claim of having produced no, or only one, child is true to the best of their knowledge and that they understand that they will be prosecuted for making a false statement.
- The claims of applicants regarding the number of children they have produced would be published online so that those familiar with the applicant's history could check the veracity of the claims. A large reward would be offered for disclosure of false claims. The granting of these rewards should be widely publicized to encourage

even more tips. Some law firms might make a profitable business of detecting fraudulent claims and collecting the reward money.

- Those convicted of making false claims about parentage would have to return all the money they obtained from selling their illicit credits plus pay a hefty fine. If they no longer had enough money for the fine, they would have to work until they paid it off. Those who used their illicitly acquired credits for reproduction would be sterilized and their illicitly produced children would be denied population credits. This would compensate for the illicit increase in the population that these children represent.

Whenever the facts of parentage are in dispute, DNA evidence would be used to establish the truth.

After 50 years, there will no longer be any need to deter or punish false claims, because false claims will be impossible. That is because the entire reproductive population will be registered by that time (because all babies will have been registered at birth).

The reason people 50 and older would not be issued population credits is that these credits are intended to regulate reproduction in order to reduce the population size to an optimal level. They are not intended as a government handout. If people 50 and older were granted credits, they would almost always sell them or donate them, and this would result in population increase (either from domestic newborns or from immigrants, depending on who bought or received the credits). Another reason for not providing credits to people over 50 is to discourage middle-aged men from producing children. These men have a relatively short working life ahead of them, have lower energy levels, and are much more likely to develop serious health problems. They are also more likely to transmit birth defects such as Down's syndrome and schizophrenia to their offspring. Moreover, when people reach middle age they often need to devote their time, labor, and money to caring for their own declining parents. Therefore, it is not in society's interest to encourage breeding by those who are past their reproductive prime.

Even though citizens older than 50 would not receive population credits, they would still be free to produce children if they are physically able to. They would simply need to acquire the necessary two population credits on their own, either receiving them as a gift or purchasing them.

Some people who are issued credits will be within a few days or weeks of their 50th birthday at the time of credit issuance, leaving them little time to dispose of their credits. Therefore, anyone who is within 30 days of turning 50 on the date they receive their credits would be granted an automatic 30-day extension in order to dispose of their credits.

Because credits would expire when people turn 50, those turning 49 should be especially diligent about using contraception, lest a baby be born after their credits have expired.

Each person's registration information would reside in his or her reproductive account, which would be part of a nationwide electronic vital statistics database. The only part of the database that would be accessible to the public would be the number of population credits each individual possesses.

Special provisions would be made for registering people who are unable to come to a registration center, such as those who are hospitalized, reside overseas, occupy mental institutions or prisons, or cannot leave home due to illness or disability. Registrars would visit those who cannot come to a registration center. Those living overseas would register at the nearest consulate.

During the initial registration period, the government and private organizations would launch a major reproductive education campaign in the schools, on television, and online, with the aim of reducing the number of unwanted pregnancies.

Although not directly connected to the population credit plan, the registration process should include a test for HIV. Those who test positive should be enrolled in a carefully supervised mandatory treatment program. The drugs currently used to combat this virus greatly reduce its transmissibility. Therefore, a program to treat all carriers (most of whom are unaware they are carriers because the disease has a latency of up to 10 years) would swiftly halt the spread of this pernicious virus (which infects 55,000 Americans every year and has killed millions worldwide). All immigrants and visitors to the country should also be screened and treated. Whenever an isolated case of HIV/AIDS subsequently turns up, the patient should be interviewed to learn his or her sex partners, so they can be contacted and tested (as has traditionally been done for other venereal diseases). It is also important to prevent and treat gonorrhea, because that disease damages the vaginal epithelium, enabling HIV to enter the bloodstream. When AIDS first emerged in the early 1980s, there was concern about protecting the privacy of victims, because they faced discrimination. Today, more enlightened attitudes prevail, so privacy concerns should not be allowed to prevent the elimination of a disease that is dangerous and expensive to treat. It would be unethical not to put an end to AIDS when we can so easily do so.

Designated Centers for
Reproductive Services

Hospitals with maternity wards would not only register newborns and adults so that they could be issued population credits, but would also provide contraception, abortion, sterilization, and genetic screening. These services would be fully federally funded. Those who object to paying taxes to support abortion would be eligible for a tax checkoff equivalent to the per capita cost of funding abortion. (This is explained in a separate chapter that follows.)

If a baby is born outside a maternity ward, the parents would have to register it within 24 hours. No birth certificate would be issued until the baby was properly registered.

National DNA Processing Laboratory

A national DNA laboratory would be established to process all DNA samples. It would be part of the national health service, not the national police. (In the United States, this would be the Department of Health and Human Services.) The handling and processing of DNA samples would include multiple safeguards to ensure accuracy and privacy. Each person's DNA information would reside in his or her account in the national vital statistics database.

An interesting consequence of accurately determining parentage is that concealment of paternity would no longer be possible. Some fathers are unaware, or prefer to ignore, that they are raising another man's child as a result of their wife's infidelity.[1] Once a population-stabilization plan is in place, the father will have confirmation of his paternity (or lack thereof). The societal consequences of exposing true paternity might include a decrease in women having extramarital sex, greater use of effective contraceptives, a slight increase in the abortion rate, and a slight increase in the divorce rate. All these outcomes would help to lower the birth rate. The legal system will need to respond to the new certainty about paternity by making the biological father financially responsible for the baby. This will further lower the birth rate by giving cheating males a strong economic incentive to use contraception. Confirmation of parentage will also reveal instances of close inbreeding. Although inbreeding can reinforce desirable traits, it also increases the likelihood of harmful double-recessive traits.[2] This is especially problematic when inbreeding is multigenerational. In any case, the consequences of mate selection should be sorted out by natural selection, not the legal system.

Just as DNA testing will reveal to husbands whether their wives have been reproductively unfaithful, wives will be able to discover whether their husbands fathered a child behind their backs. A quick check of his population credit status would reveal whether he had to spend a credit or two to pay for a clandestine baby (unless he was able to persuade his lover to pay both credits from her account).

An important service that the national DNA laboratory would offer to those considering parenthood is a genetic assessment of the likelihood of transmitting a serious genetic defect or disease to their offspring. This service would be free and would include counseling. If people knew that their offspring would likely inherit a debilitating disease or defect, most would have the good sense to forego reproduction and instead adopt. This would spare both the parents and the offspring a great deal of suffering and spare society large expenses. Providing parents with this kind of genetic information would also lower the abortion rate, because a pregnant woman would not have to wait until she was pregnant to learn that her fetus was defective and then elect to have an abortion.[3]

Notes

1. One assessment found that about 2 percent of US babies are products of infidelity. See "How Well Does Paternity Confidence Match Actual Paternity? Evidence from Worldwide Nonpaternity Rates" by K. G. Anderson, *Current Anthropology* 48, (2006).

2. If a woman breeds with a nonrelative, the likelihood that her baby will have a serious genetic defect is 1 in 50. If she has a baby by her first cousin, those odds double to 1 in 25. If she has a baby by her brother or father, the odds double once again to about 1 in 12. However, not many people have 12 children, so most babies of even the most closely related couples would be fine.

3. Techniques are rapidly being developed to identify fetal defects from fetal DNA in the mother's blood. This will eliminate the need for amniocentesis, with its attendant risk of inducing a miscarriage.

National Vital Statistics Database

Each registered person's name, DNA information, population credit status, personal identification numbers (Social Security, passport, driver's license), reproductive history, birth certificate, and (eventually) death certificate will be stored in his or her individual account within the national vital statistics database. Storing the DNA information of all Americans would require about 1 petabyte of storage space. Like the national DNA laboratory, the national vital statistics database would be the responsibility of the national health service.[1]

Members of the public would be able to query the database to learn the population credit status of anyone they are considering reproducing with. Citizens would be able to make these inquiries quickly by visiting the vital statistics website or phoning a toll-free number. They would only need the person's Social Security number, driver's license number, passport number, or (less reliably) name and birth date.

The national vital statistics database would also collect nationwide mortality data. All death certificates would be in both electronic and paper formats. The advantage of the electronic format is that death certificate information could be instantly transmitted to the vital statistics database. This precise, up-to-the-minute information about the mortality rate and cause of death would enable the national health service to promptly detect disease outbreaks, so that timely defensive measures could be taken. Having accurate mortality data would also help to keep the population size at an optimal size. If mortality went up, credits that had been retired could be released back to the public (perhaps through a public drawing); if mortality went down, more credits could be purchased by the government and retired. Computers linked to the vital statistics database could handle these tasks automatically.

In the case of babies produced by sperm bank donors, the contract information in the accounts of the mother, her husband/partner, and the sperm donor would prompt the computer database to automatically deduct the correct number of credits from the woman's and/or her husband's/partner's account(s), rather than from the sperm donor's account. This matter is covered in detail in the chapter "Special Situations."

Whenever the vital statistics database receives confirmation of the death of a person under the age of 18, the two credits belonging to the deceased would automatically be returned to the account(s) of the person(s) from whom they were originally paid. One exception would be the case of parents who have produced a third child illicitly without paying two credits. In that case, the deceased child's credits would be automatically redirected to the account of the illicitly produced child (who would not have received any credits when he or she was born). This whole procedure will be covered in detail in the chapter "Canceling the Impact of Unaccredited Births." Many aspects of the database would be automated. For instance, the database would automatically delete any remaining credits from the accounts of everyone who turns 50. This would occur at 12:00 a.m. on their 50th birthday. It would also automatically delete any unused credits from the accounts of those who die between their 18th and 50th birthdays.

Having a complete national vital statistics database (a population register) would greatly improve the accuracy of the US census. Eventually, when all the population is registered, the database will allow us to know the exact number of people in the country on a day-to-day basis as well as their precise breakdown by sex and age. This will eliminate a chronic problem of US censuses: undercounting. It will also enable better economic and social planning. The national vital statistics database would supersede the traditional national census. It would be a constantly updated census. The elimination of the traditional census would save taxpayers a great deal of money: the 2010 US census cost $13 billion, and the 2020 census cost $15.6 billion. Information such as ethnic self-identification and household information could be replaced by scientific surveys.

Every practical measure would be taken to protect the national vital statistics database from hacking and computer viruses. It would also be protected from the destructive electromagnetic fields generated by large solar flares.

Notes

1. The United States already has a National Vital Statistics System. It is part of the Department of Health and Human Services. However, unlike many other countries, the US does not maintain a population register. This is a database of all the citizens of a country with records of their birth, death, marriage, divorce, and change of residence. There are currently about 60 countries that do maintain such registers. Kuwait recently went a step farther by requiring that DNA records be kept for all residents.

Privacy Protections

Law enforcement agencies would not be granted direct access to DNA information in the national vital statistics database. However, they would be able to send DNA samples to the national DNA laboratory, which would process the DNA and forward the results to the national vital statistics database, where an automated search for a match would take place. The results would then be forwarded to the police.

The national DNA laboratory would replace all locally operated DNA crime laboratories. This would be a good thing, because many of these local laboratories are chronically underfunded and have large backlogs of samples, which seriously undermines public safety. Local laboratories are also subject to tampering by technicians eager to please prosecutors.

The DNA information in the national vital statistics database would be available to all legitimate medical researchers investigating genetic factors in disease. However, the identities of the people who contributed the DNA would be blocked from the researchers unless the individuals gave their consent.

Personnel with access to the national vital statistics database would be carefully screened before being hired and would be required to sign a lifetime agreement not to divulge any private information held in the database. This signed agreement would be comparable to similar agreements made by employees of government intelligence agencies. Periodic screening of personnel with database access would take place, using procedures equivalent to those used by the US Federal Bureau of Investigation (FBI) and the Central Intelligence Agency (CIA) to deter and uncover leakers.

The vital statistics database would hold the following information on every registered citizen:

- Legal name (and any former legal names)
- Identification numbers (Social Security, passport, driver's license)
- Sex (genetic sex and sexual identity)
- Birth date/current age (along with a copy of the birth certificate)
- Death date (along with a copy of the death certificate)
- Names and Social Security numbers of everyone the person reproduced with
- Names and Social Security numbers of all the children the person produced
- Number of population credits currently in the account
- Identity of any child produced illegally (that is, without credit payment) and the disposition of the case
- DNA data
- Sperm bank donor contract information (if applicable)

Some may protest that holding this information would be a violation of privacy, but much of this information is already held by various government agencies. And private entities such as banks, cell phone companies, internet companies, online social websites, and medical insurance companies hold vastly more information of a personal—even intimate—nature. A population-stabilization program cannot function without elementary information about people.

As noted above, each person's population credit status would be available for anyone to check. The reason for making credit status publicly available is to keep sexual partners honest. If credit status could be kept secret, then people who lacked credits could tell their sexual partner that they possessed two credits when they didn't. The other partner would only discover the deceit at the time the baby was born—too late. The deceived parent would then have to use his or her own two credits for the baby. And if the deceived parent lacked two credits (and couldn't acquire them by donation or purchase), he or she would have to choose between sterilization or leaving the country.

As mentioned in an earlier chapter, making everyone's population credit status available would also be useful for combating false claims made during the initial registration process about the number of children a person has produced. Citizens would be able to look up the credit status of people they know and collect a handsome reward if they tip off the authorities to instances of fraud. These rewards, combined with stiff penalties, would eliminate nearly all false claims regarding the number of children produced.

Deterring Illegal Immigration

For a population-stabilization program to function successfully, illegal immigration must be halted. That is because illegal immigrants often remain in their adopted country, and a large percentage of them reproduce there. These extra members of the population have social, economic, and environmental costs. Excluding illegal immigrants would not only prevent these costs but give the countries the immigrants come from a strong incentive to get serious about lowering birth rates and reforming dysfunctional social, economic, and political institutions. Currently, they can indefinitely postpone these reforms, because porous international borders serve as convenient safety valves for relieving population pressure.[1] Moreover, the immigrant-exporting countries derive great economic benefit from the cash remittances sent home by illegal immigrants, which means these countries have a strong incentive to encourage their citizens to emigrate (especially those who are poor and uneducated). In effect, these countries function like employment agencies, permanently leasing out millions of their surplus laborers to employers in more prosperous lands.

About one-half of America's roughly 11 million illegal immigrants arrived illegally. The other half arrived legally but then overstayed their visas or Border Crossing Cards. What motivated nearly all these illegal immigrants to come to the United States was the knowledge that they could get jobs that pay much more than they could make in their home countries. If these jobs were no longer available, immigrants would no longer come, and many of those already here would return home. This makes the task of ending illegal immigration relatively simple. We need only enforce the existing federal law that prohibits the hiring of illegal immigrants. There is no need to build a great wall on the Mexican border. The US government already has a database called E-Verify that employers can use to screen for illegal immigrants. But as of 2021, only 9 states (all of them in the South and Southwest) require all or nearly all employers to use E-Verify. An additional 11 states require E-Verify for public employees. In contrast, California and Illinois, which have large populations of illegal immigrants, have enacted legislation to prevent the use of E-Verify. One weakness of E-Verify is that federal legislation prevents the states from prosecuting employers who hire illegal immigrants. States are only allowed to suspend or revoke business licenses. Another weakness of E-Verify is that it fails to prevent illegal immigrants from using counterfeit identification to obtain employment. This problem could be easily fixed by requiring that Social Security cards have the same security features as passports, including a photo or other biometric data.

Implementation of the two-credit system would be accompanied by federal legislation requiring all employers to use E-Verify not only for new hires, but to screen their entire existing workforce. A deadline would be set for employers to certify that their workforce was 100 percent legal. After that, any employer found to have illegal employees would be arrested, tried, and punished. The punishment for a first offense should be a fine substantially larger than the profit made from employing the illegal immigrants. Employers convicted a second time should go to prison. To assist law enforcement, generous rewards should be given to citizens and noncitizens who report employers of illegal workers. The US should also offer financial inducements to Latin American countries to help deter illegal immigration (and should fund birth control programs there).

To deter pregnant women from entering the US to give birth (thereby enabling the baby to automatically obtain US citizenship), a federal law should be enacted declaring that the 14th Amendment of the Constitution does not grant US citizenship to babies born to foreign mothers. This would bring the US into line with other countries. Alternatively, the Supreme Court could simply rule that birth citizenship does not apply when the mother is a foreigner. A US president could trigger a Supreme Court decision by issuing an executive order declaring that in the future the United States will no longer recognize babies born to foreigners as American citizens. Instead, the birth certificate will list the country of origin of the mother.

The reason Congress has not already passed a law that would require strict enforcement of E-Verify is that members of Congress receive large campaign donations from businessmen who profit from cheap illegal immigrant labor. Most congressmen only pretend to oppose illegal immigration. And they abandon even that pretense when they vote for amnesties for illegal aliens.

Even if we had a policy of strict enforcement by E-Verify in all 50 states, we would still need the Border Patrol to deter smugglers, terrorists, and those seeking to join family members in the US. This border policing could be accomplished most efficiently and effectively if our government cultivated a network of paid informants in Mexican border towns. The informants would notify US border authorities when and where people were planning to illegally cross the border. We should also install more remote cameras and infrared sensors (rather than expensive physical barriers). These surveillance devices could also be used by biologists to gather information about wildlife movements across the border.

Under the US criminal code (Title 8, Section 1325), illegal entry of noncitizens into the US may be punished by a fine of $50 to $250 and/or imprisonment for not more than 6 months. Punishment for a second offense may be punished by a larger fine ($100 to $500) and/or imprisonment for not more than 2 years. In the past, the government has preferred to quickly hustle captured illegal immigrants back across the border rather than prosecute them. This was a sound policy for three reasons: first, prosecutions are expensive; second, they are ineffective as a deterrent (the poor have little to lose); third, the people who are most responsible for illegal immigration are not the immigrants themselves but the people who hire them (the baiters). It is these employers who should be the target of prosecutors. In 2021, the US expelled 1,000,000 illegal aliens, but only prosecuted 267—the lowest level of prosecution since 1986. Even so, most people in US federal detention continue to be illegal immigrants. This is a huge waste of taxpayer money and is both ineffective and cruel.

Illegal immigrants should be treated with compassion and respect. They are mostly good people doing the best they can to make a better life for themselves and their families. We would do the same in their place. Once they understand that they will not be able to obtain employment in the US, most of them will give up their dreams and stay home. (Exceptions would be those fleeing violence and those trying to reunite with their children and other family members.) When illegal immigrants are detained, they should be returned to their native lands as quickly as possible. If they decide to appeal their deportation, they should be lodged in decent housing in secure camps, where adult males are separated from females. They should never be put in jails unless they are violent (and in that case, they probably need mental health services). Above all, they should never be separated from their children. Under current US policy, the children of illegal immigrants can be seized from their parents when the parents are jailed for illegal entry. The children can then be placed in foster care or even put up for adoption.[2] This heartless policy existed even before 2016, but the Trump administration greatly expanded it.

The treatment of detained illegal immigrants should follow the standard Don Quixote laid down for Sancho Panza when Sancho was about to assume a governorship:

> Treat the culprit who falls under your jurisdiction as a pitiful creature, subject to the conditions of our fallen nature, and in everything that pertains to you, without doing injustice to another, display pity and clemency; for although all of God's attributes are equal, in our eyes mercy shines brighter and stands higher than justice.[3]

First-time deportees should be given enough money to cover travel expenses back to their native town or village, plus a little extra to tide them over. This travel money can come from the fines imposed on employers convicted of hiring illegal immigrants. People deported a second time should receive only bus fare home.

Until a few years ago, illegal immigration was a do-it-yourself affair: immigrants simply climbed the border fence or waded across the Rio Grande. But nowadays, Mexican criminal gangs control much illegal immigration. Even to ford the Rio Grande, immigrants must often pay a fee or agree to transport illegal drugs. They may also be beaten, robbed, raped, and abandoned to die of thirst in the desert. One of the many benefits of halting illegal immigration would be the elimination of the exploitation and abuse of immigrants by criminals in Mexico (and by employers in the United States).

The good news is that in recent years about as many illegal Mexican and Central American immigrants have been returning home annually as are entering the US. The bad news is that many more illegal immigrants are now applying for asylum, telling the US Border Patrol that they fear for their lives in their native land. After processing, most of them are released to await an asylum hearing—a process that now takes about six years! Clearly, the asylum process needs a radical overhaul.

Wealthy nations need to address the immigration problem at its root in the immigrant-exporting countries. Long-term contraception and abortion should be made available to everyone in those countries at no cost. This should be accompanied by a well-funded educational campaign to persuade women to have small families and prolong breastfeeding (which helps prevent ovulation). The poor need to hear the truth that overbreeding is the principal cause of their poverty. Otherwise, they will blame their misery on everyone but themselves, and their suffering will only increase.

> When the wages of labour are hardly sufficient to maintain two children, a man marries and has five or six. He of course finds himself miserably distressed. He accuses the insufficiency of the price of labour

to maintain a family. He accuses his parish for their tardy and sparing fulfilment of their obligation to assist him. He accuses the avarice of the rich who suffer him to want what they can so well spare. He accuses the partial and unjust institutions of society, which have awarded him an inadequate share of the produce of the earth. He accuses perhaps the dispensations of Providence, which have assigned to him a place in society so beset with unavoidable distress and dependence. In searching for objects of accusation, he never adverts to the quarter from which all his misfortunes originate. The last person that he would think of accusing is himself, on whom, in fact, the whole of the blame lies, except in so far as he has been deceived by the higher classes of society. He may perhaps wish that he had not married, because he now feels the inconveniences of it; but it never enters into his head that he can have done anything wrong. He has always been told that to raise up subjects for his king and country is a very meritorious act. He has done this, and yet is suffering for it; and it cannot but strike him as most extremely unjust and cruel in his king and country, to allow him thus to suffer, in return, for giving them that they are continually declaring that they particularly want.

Till these erroneous ideas have been corrected, and the language of nature and reason has been generally heard on the subject of population, instead of the language of error and prejudice, it cannot be said that any fair experiment has been made with the understandings of the common people; and we cannot justly accuse them of improvidence and want of industry, till they act as they do now, after it has been brought home to their comprehensions, that they are themselves the cause of their own poverty; that the means of redress are in their own hands, and in the hands of no other persons whatever; that the society in which they live, and the government which presides over it, are totally without power in this respect, and that however ardently they may desire to relieve them, and whatever attempts they may make to do so, they are really and truly unable to execute what they benevolently wish, but unjustly promise; that when the wages of labour will not maintain a family it is an incontrovertible sign that their king and country do not want more subjects, or at least that they cannot support them; that if they marry in this case, so far from fulfilling a duty to society, they are throwing a useless burden on it, at the same time that they are plunging themselves into distress; and that they are acting directly contrary to the will of God, and bringing down upon themselves various diseases, which might all, or the greater part, have been avoided, if they had attended to the repeated

admonitions which he gives, by the general laws of
nature, to every being capable of reason.

—Thomas Malthus[4]

Notes

1. Edward Abbey commented that "Mexico needs not more loans—
money that will end up in the Swiss bank accounts of *los ricos*—but
a revolution. A complete revolution, not communist, not capitalist,
but moral: a revolt against injustice, cruelty, oppression, squalor
and—most obvious—a woman's rebellion against Our Lady of
Perpetual Pregnancy" ("A San Francisco Journal" in *One Life at a
Time, Please,* Henry Holt and Company, pp. 70–71). In the years
since Abbey wrote those words, the Mexican birth rate has fallen
dramatically, but births continue to exceed deaths by almost 4:1
due to population momentum (CIA Factbook). A positive development in 2021 is that Mexico decriminalized abortion. Interestingly,
Mexicans are no longer the majority of those living illegally in the
United States. In 2017, the US had 4.95 million unauthorized
Mexican immigrants and 5.5 million unauthorized immigrants
from other countries. However, when the Mexicans are added to
the 1.9 million unauthorized immigrants from Central America,
the total is 6.85 million, so Latin Americans still make up the majority of illegal US immigrants. See https://www.pewresearch.org/
search/Key+facts+about+the+changing+U.S.+unauthorized+immi-
grant+population.

2. Thousands of children have been seized from their detained or
deported parents and are now in foster care. The parents have little
chance of ever recovering their children. What the US government
ought to do is give each of these children a passport of the nation
their parents are being repatriated to. If foreign governments are
unwilling to issue these passports, the US government should
manufacture and issue them. If the children were born in the US,
this would not annul their US citizenship, but it would enable them

to remain united with their parents. This would be much better for the children and would save the states and federal government a great deal of money.

3. Chapter 42 of *Don Quixote*. Cervantes may have had in mind James 2:13: "For judgment will be without mercy to anyone who has shown no mercy; mercy triumphs over judgment." Of course, if you think about it, "judgment will be without mercy" contradicts "mercy triumphs over judgment."

4. Donald Winch, ed., *An Essay on the Principle of Population* (Cambridge University Press, 1992), 227.

Resolving the Resident Illegal Immigrant Problem

Once the government requires all employers to screen all their employees for legal residency, the inflow of illegal immigrants will dry up. The question then becomes what to do with the roughly 11 million illegal immigrants already in the United States. Under the plan proposed in this book, illegal immigrants would be given a choice: either return home or become US citizens. If they opt for US citizenship (as most of them surely would), they would not receive any population credits (unlike other citizens). That means that if they wanted to reproduce they would first have to acquire the necessary credits on their own. If they find this requirement too onerous, they can always return to their native land and produce children there. Noncitizens who produce babies in the US without possessing the required credits would be treated exactly like citizens: they would be given the choice of leaving the country (with or without the baby) or being sterilized. In either case, the baby would receive no population credits.

At the time a credit-based population-stabilization plan goes into effect, all illegal immigrants who wish to obtain legal status as citizens would have to register. Once their identities are established, they will be issued US citizenship and a passport. A reproductive account will also be set up for them, but it will contain no population credits. These accounts will note the reason for not issuing credits as "illegal entry" or "overstayed visa or Border Crossing Card."

Some will argue that resident illegal immigrants ought to be granted population credits just like other citizens, since doing so would not prevent the population from eventually being stabilized. The rebuttal to this is that illegal immigrants have already illegitimately added themselves to the population and, in many cases, have also added progeny. Moreover, it would be unfair to confer the same rights on illegal immigrants as on legal immigrants. A final argument is that illegal immigrants can always return to their country of origin if they feel they must reproduce (though this would exacerbate the overpopulation problem in their homeland).

By offering the option of citizenship to all illegal immigrants currently residing in this country, a political and social underclass that represents about 3.3 percent of the US population would be eliminated. This would benefit the nation by reducing economic exploitation and extending the right to vote to millions who are now disenfranchised. It would also improve social cohesion. Best of all, it would eliminate the atmosphere of fear in which illegal immigrants (and their families) must now live. Their fear of deportation is akin to the fear felt by fugitive slaves in Northern states following the Supreme Court's infamous Dred Scott decision.

Although many Americans feel that all 11 million illegal immigrants should be sent back to their homelands, there are three sound reasons why this should not be done. First, it would cause tremendous hardship for the deported people (many of whom are children who were raised in the United States and are therefore more American than foreign); second, the feeble economies of Mexico and Central America are incapable of absorbing such a massive influx; third, it would be economically disruptive to this country, because it would take considerable time to recruit and train Americans to do the jobs the immigrants perform. (Recall that illegal immigrants comprise 5 percent of the American workforce.) So rather than deport our illegal aliens, we should legalize them but deny them population credits. I think most Americans would support such a compromise. Hard-nosed conservatives would like the fact that illegal immigrants would be forbidden to reproduce (unless they legitimately acquire credits on their own), while soft-hearted liberals would be relieved to know that immigrants would be spared deportation into conditions of greater poverty and perhaps violence.

Curbing Unaccredited Births

For a population-stabilization plan to function properly, not only does illegal immigration have to be halted, but couples need to be deterred from producing a child when they lack two credits to pay for one.

Ideally, everyone would recognize the value of working toward an optimal population size and would conscientiously do their part to comply with the requirements of a credit-based population program. They would always conduct a check of their partner's credit status before engaging in unprotected sex that could lead to pregnancy. If this credit check revealed that the couple would not have the necessary two credits, they would use effective contraception or abstain from sex altogether. If the couple failed to use contraception (perhaps because of overindulgence in alcohol) or if the contraception itself failed (perhaps the condom slipped), the woman would avail herself of emergency contraception—either the morning-after pill or the copper IUD (the latter is 100 percent effective). If she failed to use emergency contraception within the prescribed period (perhaps because her partner falsely told her he had a vasectomy), she could then obtain a free chemical or surgical abortion. If her principles forbade abortion, she and her partner could try to purchase two credits before the baby's birth. The couple might acquire the money for this purchase by selling their home, by using their retirement savings, or by obtaining a loan from a bank or a relative. Perhaps abortion-prohibitionist religions would establish a loan fund to help them out. The couple might also be able to beg two credits from a relative or friend. But if they ended up being unable to pay two credits, the following would occur when the baby was born:

First, an independent laboratory would double-check the DNA of the mother, father, and baby to ensure that no mistake was made in determining parentage. The parents' credit status in the vital statistics database would also be rechecked. Parents would be allowed to dispute the accuracy of the findings. A swift and unbiased appeals process would be available. Once the lack of two credits was confirmed, the parents would have to decide whether to leave the country or be sterilized. If they opted to leave the country, they could take the baby with them or leave it behind for adoption. If they opted for sterilization, the choice of surgical procedure would be up to them as long as it was effective. Most women would choose either laparoscopic tubal ligation or suprapubic minilaparotomy ("minilap"). Unfortunately, these procedures are not 100 percent effective. It is estimated that laparoscopic tubal ligation has a 2.4 to 3 percent failure rate, and the failure rate for minilap is slightly higher.[1] About 40 percent of these failures are due to tuboperitoneal fistula, 20 percent to surgical error, 10 percent to ectopic pregnancy.[2] The fact that 20 percent are due to surgical error means that only highly qualified surgeons should be allowed to perform sterilizations.

There is a nonsurgical alternative to tubal ligation called Essure micro inserts. This involves inserting little springs into the fallopian tubes, which cause scar tissue to quickly form and block the tubes. These springs can be inserted during an office visit. However, Essure micro inserts are not as reliable as tubal ligation: it is estimated that they have a 9.6 percent failure rate, which is much higher than for laparoscopic tubal ligation.[3]

Those who think that sterilization or exile are excessively harsh sanctions ought to compare them to nature's alternatives: famine, pestilence, war, and genocide. The purpose of sterilization is to prevent the possibility of future violations by the offender, and to deter others from violating. It is not a punishment, because there is no element of retribution. It is a purely pragmatic measure. Moreover, those who opt for sterilization will not suffer any disruption to their daily lives. Everyone needs to realize that having a baby without possessing enough credits to pay for it represents stealing from everyone else. It deprives other people and other species of resources that rightfully belong to them and their descendants. Like all theft, illicit reproduction must be deterred.

When a man learns that he has impregnated a woman and that he will receive a vasectomy when she gives birth due to his inability to pay two credits, he may try to persuade the woman to have an abortion. He is certainly within his rights to make such a request, but he may not harass or intimidate the woman. If he attempts to do so, a restraining order will be promptly issued and strictly enforced with a GPS tether.

US parents who leave the country to avoid sterilization would not be allowed to return unless they agreed to be sterilized. However, the sterilization requirement would be waived if they could demonstrate that they were no longer fertile due to advanced age, hysterectomy, orchiectomy, vasectomy, tubal ligation, or other proof of sterility. If they produced additional children after leaving the United States, those children could not enter the US until the parents acquired sufficient credits for them. Until the Constitution is amended or the

Supreme Court rules otherwise, an unaccredited child born in the US who was removed by its parents could still return at any time, because the child would be a US citizen.

Relatively few native parents who are unable to pay credits will choose emigration over sterilization, because most will lack dual citizenship or a visa that would allow them to reside and work in another country. However, immigrants (whether legal and illegal) almost always have the option to return home, and they now comprise nearly 15 percent of the US population. Likewise, Orthodox Jews—a group that produces large families—can always go to Israel (though Israel is already grossly overpopulated).

Notes

1. *Chicago Tribune*, September 2, 1992.

2. Shilpa Date, et al., "Female Sterilization Failure: Review over a Decade and Its Clinicopathological Correlation," *Int J Appl Basic Med Res* 4 (July–December 2014): 81–85. (www.ncbi.nlm.nih.gov/pmc/articles/PMC4137647)

3. *Chicago Tribune*, April 21, 2014.

Canceling the Impact of Unaccredited Births

If the parents are unable to pay credits when their baby is born (a problem strictly of their own making), then no credits will be issued to the baby's account. In time, this will cancel out the illicit increase in the population that the baby represents. Although it may seem unfair to deny credits to the offspring when the fault belongs with the parents, doing so has the benefit of allowing the parents to keep the baby. A less desirable alternative would be to remove the baby for adoption and then require the adopting parents to pay the two credits.

The fact that these illicit babies would not be issued population credits does not mean that they could not have children when they grow up. They could produce one child by using the two credits of their spouse/partner. And if they want more children, they could save money and purchase credits—in effect, purchasing the birth-right that their irresponsible parents deprived them of. A couple who lacked two credits could also adopt a baby whose credits had already been paid by the birth parents. Another option would be for the woman to be implanted with an embryo for whom the credits had already been paid by the sperm and egg donors—a process known as embryo adoption or embryo donation.

If the parents of a newborn possessed only one credit to pay for the baby, that one credit would be deposited in the newborn's account. That way, when the baby grew up it would only need to acquire one additional credit to have the full complement of two.

If one of the two older siblings of an illicitly produced child dies before the age of 18, the deceased sibling's two credits would automatically be transferred to the account of the illicitly produced child (rather than being returned to the parents). If *both* older siblings died before reaching 18, one pair would be retired and the other would go to the illicitly produced child.

If an illicitly produced child, upon growing up, produced a child without having enough credits to pay for it, society would now be owed four credits: two for the original illicit child and two for the illicit child's illicit child. At that point, I think both the parents and child should be sterilized so that this pattern of illicit conduct cannot continue. Otherwise, natural selection would immediately begin favoring the violating lineage, which would weaken the two-credit system and eventually break it down. Like water trying to work its way around a dam, natural selection is always trying to work around barriers to reproduction. We cannot afford to let it succeed.

How Much Immigration to Allow

In the first edition of this book, I recommended a free market for population credits, with foreigners and natives free to bid against each other for credits. I have since shifted to a more nationalistic viewpoint, because there are so many wealthy foreigners who would like to have US citizenship that I suspect they would outbid Americans. This would increase immigration and prevent many natives who hoped to have three or more children from being able to do so. So I now believe there should be an annual quota on the number of credits available for purchase by foreigners. My own preference would be to set the quota equal to the number of people who permanently leave the US each year. So, if 100,000 Americans move out, then 100,000 foreigners could move in. Not only would this improve national cohesion, but it would substantially lower credit prices by preventing foreigners from bidding prices up and thereby preventing Americans who want to have more than two children from being able to have them.

Credit Donations and Sales

Donations

People could donate their credits to fellow citizens or foreigners. These donations would need to be officially registered. Before foreigners could receive credits, they would have to open a credit account at the nearest US consulate. The donated credits would then be deposited into their account.

Private Credit Sales Would Be Treated as Compensated Donations

People who wish to pass their credits on to a friend or relative in exchange for some kind of compensation could work out their own private arrangements (using a legal contract if they wish). Exchanging one's credits for a gift or sum of money less than the market value of the credits would still be treated as a donation for purposes of official record keeping. These compensated donations could be made to foreigners as well as fellow citizens. However, it would be illegal to advertise credits for sale, so most people would still sell their credits through the government-regulated online market.

Sales through the Government-Regulated Online Market

Citizens who want to obtain the most money for their credits would sell them through the government-regulated online market. This market would be open to people anywhere in the world who have internet access. The most advanced security software would be used to prevent cyberattacks and fraud.

Credit purchases would be limited to individuals. Private organizations, religious organizations, and corporations would be excluded. It would be illegal for such organizations to hand out money to purchase credits, and individuals would not be allowed to purchase more than 10 credits during their lifetime.

Rich people who wanted to have three or more children would have an obvious advantage over poor people. However, the rich can afford extra children, while the poor cannot. And most rich people prefer to have small families anyway. The existence of a 10-credit lifetime limit on credit purchases would prevent rich couples from having more than 12 children (unless they received additional credit donations). The first two children would be paid for with the couple's birthright credits and the next ten would be paid for with the 10 credits each spouse would purchase.

Because the government would regulate the credit market, there would be no private speculation. Speculators could not buy large quantities of credits when prices were low and sell them when prices were high because individuals would be limited to a lifetime maximum of 10 credits.

The government would purchase credits under two circumstances:

1. To reduce population size at a faster rate than would be possible by credit attrition alone
2. To prevent population increase in the event of a medical or genetic breakthrough that significantly lowered the mortality rate.

The government could also put retired credits back into circulation if there were a sudden increase in mortality, as might be caused by a lethal epidemic. This could be accomplished through a public drawing. Since all population credits would be identical (the only difference being price), whoever offered a credit for the lowest price would be the first to sell. If fewer citizens offered their credits for sale (reducing the supply), the price would rise. A rising price would encourage more citizens to forego reproduction and sell their credits, which would lower the price.

As mentioned previously, no one would ever have to sell their credits to obtain life's necessities, because poverty itself would be eliminated when the two-credit system is implemented (as described in the chapter "Eliminating Poverty").

Americans who sell their credits could use the money to start a business, buy a house, pay for a college education, invest in the stock market, pay off debts, or simply enjoy an easier life. By selling one or two credits, people of modest means could substantially improve their economic well-being.

To prevent misunderstandings, citizens who want to sell their credits would first have to sign a document acknowledging that they fully understand that by selling both of their credits they may be foreclosing their opportunity to have children. It is important that they understand that once they have sold their credits, the only way they could have children is if someone donates credits to them, or they have sufficient money to purchase credits, or they adopt a child for whom credits have already been paid by the biological parents, or they adopt an embryo whose credits were paid by the egg and sperm donors.

Credit Returns

If a child dies before the age of 18, the credit(s) paid by the parent(s) for the child would be returned to whichever account(s) they were withdrawn from. The only restriction being that the parent(s) must be less than 50 years old on the date the child died (since credits expire when people turn 50). If one or both parents die before the child dies, the credits will be permanently retired.

When a legally produced child belonging to parents who have also produced an illicit child dies, the two credits that belonged to the legally produced child will be returned not to the parents but to the account of the illegally produced (and therefore creditless) sibling, allowing that sibling to produce two children (provided his or her partner also has two credits) or otherwise use the credits.

Credits returned to the parents of a deceased child could only be used for reproduction or donation. They could not be sold on the credit market. This is to eliminate a motive for evil parents to kill their child in order to profit from selling the returned credits.

If immigrants who receive one or both of their credits as a donation die before the age of 18 or decide to renounce their US citizenship, the person who donated the credits could reclaim them, provided the donor was less than 50 years old on the day the immigrant died or renounced citizenship. The renunciation of US citizenship would need to be registered in the vital statistics database.

Returned credits could only be used for reproduction or donated to another prospective immigrant. They could not be sold in the credit market, in order to eliminate a motive for foul play.

Twelve-Credit Donation Limit

No more than 12 credits could be donated, because that is the maximum number that an individual could ever possess (2 birthright credits plus 10 purchased credits). This will prevent wealthy citizens from purchasing numerous credits and donating them to members of their ethnic, religious, or political group in order to increase that group's representation in the population.

Special Situations

Surrogate Mothers

If a woman gives birth to another couple's baby that was implanted in her womb as an embryo, the couple that was genetically responsible for the baby would be responsible for paying the credits. It wouldn't matter who carried the baby to term.

Fertility Drugs and Multiple Births

Multiple births are undesirable for multiple reasons: First, they are not what the parents wanted (or can afford). Second, the babies are often born prematurely and require expensive hospital care. Third, they tend to have serious health problems later in life. In the past, fertility treatments often resulted in multiple births, but recent advances in fertility treatment can now prevent multiple births.

Twins are produced in 1 out of 89 births and triplets in 1 out of 7,921 births. If the parents of twins had no previous children, they could use their four credits to pay for both babies. If they had only two credits and could not acquire additional ones, the second twin to emerge would not be given population credits. The parents, however, would not be sterilized, since the second baby was not their fault. Triplets would be handled similarly: if the parents had four credits, the first two babies would be paid for, but the third to emerge would receive no population credits.

The US has an abnormally high incidence of multiple births, and those not due to fertility treatments may be due to a diet rich in meat and dairy products that have high levels of animal hormones. Evidence for this is that meat eaters are five times more likely to produce twins than vegans.[1]

Sperm Banks

The two-credit system would not affect the ability of women to utilize sperm banks in order to become pregnant. Sperm donors would not need to pay any credits for the pregnancies that result from their sperm. Instead, the woman and her partner (if she has one) would pay the credits. If the woman has no partner, she would be responsible for paying both credits. This would generally limit her to having just one child.

Some women might choose to have one baby by their partner's sperm and another by the preserved sperm of a man who possesses a different set of desirable traits. This could be a sound genetic strategy but would require that the woman's male partner be willing to donate a credit from his account to pay for the sperm donor's baby.

Before receiving sperm from a sperm bank, the woman and the sperm donor would have to sign a contract. They would not have to meet in person to do this. The contract would specify that if a baby results from the sperm, the sperm donor need not pay any credits. If the woman and her husband/partner plan on each paying one credit for the baby, the husband/partner would also have to sign the contract, giving his/her consent to this arrangement. These contracts would be authenticated by one of the official registrars who register new babies. This contract, in electronic form, would be filed in the accounts of the woman, her husband/partner (if she has one), and the sperm donor in the vital statistics database. That way, when the baby was born and its DNA was checked against the putative parents, the contract information in the accounts of the mother, her husband/partner, and the sperm donor would prompt the database program to debit the credits from the proper account(s).

It would be illegal for a woman to donate leftover semen from a sperm donor to another woman unless the second woman first goes through the sperm donor contract procedure.

The children produced by sperm donors would be legally entitled to learn the identity of their biological father and be able to contact him if they wish. Men who do not wish to acknowledge their progeny would not be allowed to donate sperm.

Pregnancy from Rape

A woman who becomes pregnant through rape should immediately report the crime to the police, see a doctor, and take the morning-after pill (plan B) or use the copper IUD. If she fails to use postcoital contraception and becomes pregnant, she will still have the option of receiving a free chemical or surgical abortion. If she rejects abortion, two credits will need to be paid when the baby is born. If the rapist (who will be promptly identified through his DNA once everyone is registered in the system) has two credits in his account, they will automatically be deducted to pay for the baby. If he has only one credit, the woman will also have to pay one credit. If the rapist has no credits, then the woman will have to pay both credits.

In any case, the rapist will be chemically or surgically castrated in order to reduce his testosterone to a pre-pubertal level.[2] The choice of procedure (chemical or surgical) will be up to him. If two credits cannot be paid, the woman will have to choose either sterilization or exile. If she chooses exile, she can take the baby with her or leave it behind for adoption. Although it may seem harsh to sterilize or exile the woman, since her pregnancy was involuntary, the choice to give birth was hers, since she could have prevented that outcome by using plan B or the copper IUD or getting an abortion. If we fail to demand sterilization or exile in such cases, any woman who produced a child without enough credits to pay for it could claim rape and thereby avoid sterilization—especially since claims of rape are often hard to verify.

If a raped woman could not pay the credits for the baby, the baby would receive no credits. This would compensate for the illicit increase in the population represented by the baby. And it would have the additional societal benefit of diminishing the frequency of any genes that may have predisposed the father to rape.

The two-credit system would nearly eliminate rape by strangers, because the rapist would be swiftly identified by his DNA. Once identified and convicted, he would be chemically or surgically castrated, thereby eliminating his potential for further sexual predation.

Adoption

The adoption of a child would require the payment of two credits unless the credits had already been paid by the biological parents. The adoption of an embryo would likewise require the payment of two credits. The adoption of an adult citizen by another adult citizen would not require the payment of any credits.

Notes

1. See https://nutritionfacts.org/video/why-do-vegan-women-have-5x-fewer-twins/.

2. Castrates enjoy one great benefit over noncastrates: they live an average of 13.6 years longer. Castration is, by far, the most effective way for men to maximize life span. See page 257 of *Ever since Adam and Eve* by Malcolm Potts and Robert Short. Another benefit enjoyed by castrates is that they never go bald if castrated as young men.

Universal Reproductive Education

Reproductive education is essential for people to effectively regulate their fertility. Education about sex and methods of fertility regulation should begin in grade school and continue through high school. The curriculum should include accurate information about all contraceptive and abortion options, and training on how to avoid, escape, and report sexual exploitation. Because each of America's diverse immigrant and ethnic cultures has its own traditions about how males and females should relate to one another, the establishment of a national etiquette for sexual relationships would be worthwhile. Clear communication would prevent the feelings of humiliation and rage that can lead to violence.[1]

Notes

1. See "Straight People Need Better Rules for Sex" by Christine Emba, *New York Times* Guest Essay, April 7, 2022 (https://www.nytimes.com/2022/04/07/opinion/sex-consent-dating-boundaries.html2).

Universal Access to Contraception and Abortion

Under the two-credit system, every pharmacy in the nation would be required to stock all major types of contraceptives including the morning-after pill (plan B). They would also stock abortifacients. (More than half of US abortions are now pill induced.) And all health care providers would be required to make IUDs available to their patients. It would be illegal for a pharmacist to refuse to provide contraceptives or abortifacients. Citizens who are too embarrassed to purchase contraceptives or abortion pills in person could order them online or by mail order. The postal service would deliver them in unmarked packages. A full range of contraceptive, sterilization, and abortion options would be available at every hospital that has a maternity ward. Instruction on the use of contraceptives would also be available.

There won't be much demand for abortion once reproductive education becomes universal and effective long-term contraception becomes available to all at a low cost. However, anytime a pregnancy endangers the mother's life, or the fetus is badly defective, or the pregnancy stems from rape, abortion would be the proper recourse.

Contraceptives, sterilizations, abortifacients, and abortions would be entirely paid by the government to ensure that all citizens have access to them. A tax checkoff, as described in the next chapter, would be available for those who oppose abortion, so that they would not have to pay for something they consider immoral. The government should also promote breastfeeding, since lactation is a 98-percent-effective contraceptive that helps keeps births safely spaced and keeps babies healthy. Once the milk teeth erupt, infants can begin receiving some solid food, but nursing should ideally continue for 3-4 years (as is the universal practice in traditional societies). Prolonged breastfeeding would also reduce maternal mortality due to fewer lifetime births per woman. And the breastfed babies would benefit from superior nutrition, stronger immune systems, and greater access to resources (thanks to having fewer siblings).[1]

The cost to the government of providing free contraception and abortion would be more than compensated by the cost savings from fewer childbirths. In 2010, two-thirds (68 percent) of the 1.5 million unplanned births that occurred in the US were paid by public insurance programs, primarily Medicaid.[2] By comparison, 51 percent of births overall (planned plus unplanned) and 38 percent of planned births were funded by these programs. In 2010 alone, total government expenditures on unwanted pregnancies amounted to $21.0 billion—$14.6 billion by the federal government and $6.4 billion by state governments. In 19 US states, public expenditures related to unintended pregnancies exceeded $400 billion in 2010. In that year, Texas spent $2.9 billion, California spent $1.8 billion, New York spent $1.5 billion, and Florida spent $1.3 billion.[3]

Individual US states have no legitimate role to play in regulating population size. Only the federal government has the legal power to regulate immigration, and the states have been forbidden to outlaw contraception by a US Supreme Court decision, *Griswold v. Connecticut*, 1965. Now that the Supreme Court has rejected the legal principle of stare decisis by overturning *Roe v. Wade* (the 1973 ruling that legalized abortion), the US should adopt a federal law that will make the regulation of population *explicitly* a federal responsibility—a law that will guarantee all citizens free and easy access to reproductive education, contraception, sterilization, and abortion.

Notes

1. Both the practice of wet-nursing and the direct sale of breast milk have interesting economic, genetic, and demographic consequences. They allow wealthy women (who can afford wet nurses) to have many more children than they otherwise could, while simultaneously reducing the number of children that the milk producers can produce (since lactation suppresses ovulation). This satisfies the selfish genes of both parties. The woman of means can produce many more copies of her genes, while the poor woman benefits from the additional income provided by selling her milk and the greater spacing of her births, which improves the likelihood that her own children will survive, thrive, and reproduce. The poor woman is also less likely to die in childbirth, since she produces fewer children during her reproductive years.

2. A. Sonfield and K. Kost, "Public Costs from Unintended Pregnancies and the Role of Public Insurance Programs in Paying for Pregnancy-Related Care: National and State Estimates for 2010," Guttmacher Institute, 2015, http:// www.guttmacher.org/pubs/public-costs-of-UP-2010.pdf.

3. See https://www.guttmacher.org/fact-sheet/unintended-pregnancy-united-states#15a.

On Abortion

Regardless of where one stands on abortion legalization, there are three things nearly everyone can agree on:

1. We would rather follow our conscience than be dictated to.
2. We would rather live than die (even suicides would rather live but see death as the only escape from their pain).
3. We would rather avoid an unwanted pregnancy than get an abortion.

If we agree on these things, then we are all pro-choice, pro-life, and antiabortion. In other words, the terms pro-choice, pro-life, and antiabortion are "void for vagueness" (a legal term) and should be thrown out. They are mere propaganda. As honest replacements, I propose we adopt *abortion criminalizer* for those who wish to criminalize abortion and *abortion legalizer* for those who favor legalization.

Abortion criminalizers have a single goal: to reduce the death rate of the preborn by forcing all pregnant women to give birth. Abortion legalizers, on the other hand, have two goals: first, to ensure the freedom of every woman to decide whether and when to give birth; second, to protect the lives of those who have already been born. There are several ways in which legal abortion reduces postnatal death rates:

1. Legal abortion reduces the death rate of pregnant women who, if abortion were illegal, would often resort to life-threatening illegal abortions. In Chile, to cite one example, approximately 38 percent of total maternal deaths are due to illegal abortions (and this figure only includes women who died in hospitals).[1]

2. Legal abortion reduces deaths from birth complications. The act of giving birth in the United States is 15 times more likely to result in the mother's death than obtaining a legal abortion.[2] The US maternal mortality rate for childbirth in 2016 was 17.4 deaths per 100,000 births,[3] and the US had 623,471 abortions that year.[4] So if all the women who got abortions in 2016 had instead been forced to bear children, 108 of them would have died. Over the course of a decade, the maternal death toll would exceed 1,000.

3. Legal abortion reduces the murder rate three ways. First, it eliminates the incentive for infanticide by preventing the existence of unwanted infants.[5] Second, by eliminating the existence of unwanted children, it reduces the likelihood that they will be abused and grow up angry and primed to kill.[6] Third, it eliminates the incentive for a man to kill a former girlfriend he has impregnated in order to escape the financial and personal burdens of fatherhood.[7]

4. By reducing population growth, legal abortion reduces deaths due to climate change. Our emissions of carbon dioxide, methane, and nitrous oxide are already causing unprecedented droughts, heat waves, fires, floods, and catastrophic hurricanes. In an overpopulated world, giving birth causes a net increase in the premature deaths of those already born, while abortion reduces such deaths. It is abortion that is pro-life.

5. By slowing the growth of human populations, abortion slows the rate at which plant and animal habitats are overrun by humans. This preserves nonhuman lives, protecting biodiversity and keeping ecosystems healthy.

Yet another way legal abortion reduces harm is by eliminating the abandonment of infants by parents too poor to care for them. When the former dictator of Romania, Nicolae Ceaușescu, outlawed abortion for all women except those who already had at least five children, thousands of unwanted infants were given up to orphanages, where they suffered neglect and malnourishment. This led to long-term mental, emotional, and physical impairment and suffering.

The children of women who are forced to give birth often develop psychological problems. Malcolm Potts and Thomas Hayden in their book *Sex and War* (pp. 94–95) cite a multidecade study of Czech children that compared 220 children whose mothers had been denied abortion to 220 children from the same communities whose mothers had not sought an abortion. The children were otherwise evenly matched. The researchers found that the children of those who had been denied abortion "felt less positive about themselves, were rejected by friends more often, and were less likely to perceive themselves as happy. By their early twenties, the 'unwanted' children had more problems with alcohol and were twice as likely to end up in prison."

Many abortion criminalizers belong to religious sects that oppose sex education, effective contraception ("abstinence only"), and easy access to the morning-after pill. By opposing measures that would prevent unwanted pregnancies, they create demand for abortion, and thereby ensure that abortion will continue—legally or illegally. They might as well chase their shadows.

With equal lack of logic, many abortion criminalizers believe that God shares their opposition to abortion. But if God opposed abortion, there would be no miscarriages or stillbirths. The fact that 15 to 20 percent of recognized pregnancies terminate in miscarriages and stillbirths ought to convince any believer that God is an enthusiastic abortionist.[8]

At the core of religious opposition to abortion is the belief that God implants a soul in every human zygote and that all souls have equal value.[9] This has led abortion criminalizers to conclude that a zygote is as valuable as an adult and deserves equal legal protection. What they ignore is that the transformation of a zygote into an adult requires large inputs of energy, matter, and time—inputs that make the adult far more valuable. Among other things, these inputs enable adults to create many zygotes, but a zygote can never create an adult (just grow into one). Although a human zygote and a human adult both have human DNA, and both have the spirit of life, the adult is worth far more. That is why, when times are hard, nature sacrifices the young first (especially embryos) and preserves reproductive adults for as long as possible. Even abortion criminalizers, if forced to choose between killing a zygote or killing an adult, would—one hopes—spare the real person rather than the "personhood" of the zygote. When put to it, they do know the difference.

Of course, if we lived in an optimally populated world, we would try to reduce demand for abortion as much as possible. After all, abortion is unpleasant, costly, and occasionally results in medical complications. There are five simple measures that would greatly reduce demand for abortion: (1) educate young people about sex, reproduction, and contraception (and forewarn them how easily sexual desire can overwhelm reason and prudence, especially under the influence of alcohol),[10] (2) make the morning-after pill and long-term contraception readily available at low (or no) cost, and encourage their use (perhaps by offering a monetary reward to young women who are not financially or mentally ready for motherhood), (3) make it socially unacceptable to use serial abortion as a substitute for long-term contraception, (4) provide couples with free genetic screening in order to forewarn them about any genetic problems they might pass on if they conceived, (5) equalize economic opportunities for both sexes so that parents are not inclined to abort female fetuses (as commonly happens in China and India).[11] If we took these five simple measures, we could nearly eliminate abortion, and both abortion criminalizers and abortion legalizers would be happy.

But even if we took all the above measures, abortion would still be necessary in some cases. It would still be the proper recourse whenever a woman's life was endangered by her pregnancy. Abortion would also be the right choice whenever the fetus was severely defective and destined to have a wretched life (or the medical expenses of keeping it alive would deprive others, with better prospects, of good care). Abortion would also be the right course whenever the pregnancy was the result of rape. A mother cannot help but have ambivalent or hostile feelings toward a child of rape, and those feelings will be picked up by the child, with unhappy consequences. Moreover, a child of rape carries whatever genes may have predisposed the father to sexual aggression.

In thinking about the ethics of abortion, it is worth keeping in mind that life is always accompanied by many hardships. These hardships increase toward the end of life, when people suffer painful cancers, diabetic amputations, colostomies, kidney dialysis, macular degeneration, hearing loss, dementia, depression, emphysema, aphasia from strokes, crippling arthritis, broken hips, incontinence, Parkinson's disease, congestive heart failure, ALS, and a welter of other afflictions. Our lives begin with a bang and end with a whimper.

> When nature, after struggle tears the child
> Out of its mother's womb to the shores of light,
> He lies there naked, lacking everything,
> Like a sailor driven wave-battered to some coast,
> And the poor thing fills all the air
> With lamentation—but that's only right
> In view of all the griefs that lie ahead
> Along his way through life.
>
> —Lucretius[12]

In contrast to the hardships endured by those who live outside the womb, a fetus that will be aborted enjoys a cozy existence followed by a swift death. It never suffers hardship. Nor does it cause hardship for others (except for the mother). And if, as many abortion criminalizers believe, dead fetuses go to heaven (or at least to a benign limbo), that should be cause for celebration, not condemnation.

Women are the best judges of their mental, physical, and financial resources and their aspirations. They alone are qualified to decide on abortion. Male politicians and clerics should focus on eliminating the abuse of women and children, including rape and genital mutilation (male and female). And they should ensure that contraception and abortion are available to everyone at no cost (as they are now in France).

> So then, each of us will be accountable to God. Let us therefore no longer pass judgment on one another, but resolve instead never to put a stumbling block or hindrance in the way of another.
> —Paul the Apostle, Romans 14:12–14

Notes

1. See B. Viel, "Illegal Abortion in Latin America," IPPF Med Bull 16, no. 4 (August 1982): 1–2.

2. In the US, the pregnancy-associated mortality rate among women who deliver live neonates is 8.8 deaths per 100,000 live births, whereas the mortality rate related to induced abortion is 0.6 deaths per 100,000 abortions. See https://www.ncbi.nlm.nih.gov/pubmed/22270271.

3. See https://www.usnews.com/news/healthiest-communities/articles/2020-01-30/why-the-new-us-maternal-mortality-rate-is-important.

4. See https://www.cdc.gov/reproductivehealth/data_stats/abortion.htm.

5. Between 1970 and 2000, the US infanticide rate increased from 4.3 per 100,000 to 9.2, and then declined to 7.2 in 2013. These

infanticide figures would inevitably be higher if abortion were unavailable, and they would be lower if every woman had easy access to affordable contraception and abortion and received proper instruction on how to care for infants. Most Americans are unaware that, until recently, infanticide was a common practice throughout the world. In cultures where infanticide was practiced, an unwanted infant would usually be deposited in a public space (such as a marketplace) where anyone could take it. Since there were few takers, most of the abandoned infants soon perished of hunger and exposure. In India and Pakistan, infanticide is still a common practice. It was also common in China up to and including the one-child era. In Europe, infanticide was common until the Church established foundling hospitals in the Middle Ages. But those hospitals merely prolonged the suffering of the infants, since nearly all of them soon died. In the Dublin Foundling Hospital between 1775 and 1796, 10,277 infants died—a death rate of 99.6 percent.

An interesting fact about infanticide is that it is far more likely to be committed by stepfathers than biological fathers. Natural selection favors this, because the death of the infant causes the mother to stop lactating and resume ovulating. That enables the stepfather to sire his own children.

6.   In the United States, crime rates have dropped about 50 percent in the past three decades. About 45 percent of the decline may be due to the legalization of abortion by the Supreme Court in 1973. The first study to conclude that legalizing abortion reduced crime was conducted by John Donohue and Steven Levitt and published in 2002. (See Freakonomics Radio, episode 384, "Abortion and Crime, Revisited" at www.freakonomics.com.) The authors published a follow-up study in 2019 that bolstered their original conclusions. Their second study was called "The Impact of Legalized Abortion on Crime over the Last Two Decades" (National Bureau of Economic Research, May 2019).

Another important factor in crime reduction has undoubtedly been the reduction in teen pregnancies due to higher rates of both celibacy and contraceptive use. It is well established that the unwanted children of teenagers are particularly vulnerable to neglect and abuse. See https://journals.sagepub.com/doi/full/10.1177/001112871561588: "Crime, Teenage Abortion, and Unwantedness" by Gary L. Shoesmith in *Crime and Delinquency*, November 2015, vol. 63, issue 11, pp. 1458-90.

7. The protagonist of Theodore Dreiser's Pulitzer Prize–winning novel *An American Tragedy* was driven to murder his former girlfriend after all his attempts (and his girlfriend's attempts) to procure an abortion were unsuccessful, whereupon the girl insisted upon marriage and threatened to expose him if he refused. He dreaded such exposure because it would mean the loss of his job, the loss of his social standing, and the loss of another woman he had fallen in love with. The young man was duly convicted of murder and executed, and thus two young lives were senselessly extinguished that would have been saved if abortion were legally available.

8. Many miscarriages are unrecognized because they occur before the woman even realizes she is pregnant. By convention, spontaneous deaths of the unborn are called miscarriages if the fetus is less than 5 months old, and stillbirths if older.

9. The Roman poet Lucretius mocked the notion of a spirit (or spirits) waiting around to enter a new being:

It seems more than a trifle comical
To think that spirits come around in throngs
As stand-bys at the copulating rites
Or births of animals, and all agog
To be the first aboard; perhaps they have
Some mutual agreement, or each holds
A ticket for his place in line, to keep them
From scuffles, squabbles, and unseemly jostling!

Lucretius, The *Way Things Are: The De Rerum Natura of Titus Lucretius Carus,* trans. Rolfe Humphries (Bloomington, IN: Indiana University Press, 1968), 108.

10. When people are under the influence of a strong emotion, they behave in irrational ways. For example, most women, when not sexually aroused, believe they would refuse to have sex with a man who refused to use a condom. Yet women commonly have unprotected sex when aroused. Likewise, most men, when not sexually aroused, would deny that they would rape their girlfriend. Yet when sexually aroused, such rapes are not uncommon. And most women, in a normal emotional state, are confident they would respond with righteous indignation if sexually harassed. Yet when actually harassed, nearly all women clam up in fear. One lesson to be drawn from this is that a sexually active woman should always use *long-term* contraception (such as an IUD), and always keep a condom handy to prevent disease. And, of course, men should behave with equal responsibility. For more about the ways arousal distorts behavior, listen to "The Hot-Cold Empathy Gap" on the National Public Radio program "Hidden Brain," August 2, 2020.

11. When abortion is used to skew the sex ratio in favor of males, it also skews the balance of estrogen and testosterone in society. The extra testosterone makes the society more prone to civil unrest and war. Moreover, many young men will be unable to find wives, which further contributes to unrest.

12. Lucretius, *Lucretius: The Way Things Are: The De Rerum Natura of Titus Lucretius Carus*, trans. Rolfe Humphries (Bloomington, IN: University of Indiana Press, 1968), 165–166. Author Malcolm Potts observed that "we may with some truth say that even life itself is a fatal sexually transmitted disease" (*Ever since Adam and Eve*, p. 255). That echoes the Buddha's observation that death is not caused by disease or violence, but by birth.

Tax Checkoff for Opponents of Abortion

In order to accommodate citizens who have ethical objections to abortion, federal tax returns would give people the option of diverting the per capita federal subsidy for abortion (including the abortifacient RU-486) to a special fund for children who have been severely abused by parents or guardians.

Thanks to this abortion tax checkoff, all federal costs for abortion would be borne exclusively by those who have no moral objection to abortion. People who oppose abortion could pay their taxes with a clear conscience. The level of abortion service would not be affected, because the taxes contributed by those who support abortion would more than suffice to cover all costs. Also, with universal reproductive education and universal availability of contraception, the demand for abortion would be much lower than at present.

The Copper IUD
vs. Oral Contraceptives

The copper IUD (brand name Paragard) is an inexpensive contraceptive that is more than 99 percent effective (vs. 93 percent for the Pill).[1] The copper IUD has few side effects, requires no maintenance, remains effective for up to 10 years, and can be removed at any time. It also allows women to ovulate naturally, thereby preserving the benefits of a natural hormone cycle. Another desirable feature of the copper IUD is that it will prevent pregnancy if inserted within 5 days after sex.

Birth control pills, in contrast, are expensive, require women to remember to take them daily (causing forgetful women to become pregnant),[1] sometimes cause blood clots in legs or lungs (0.3 to 1 percent of women over the course of 10 years), and disrupt the natural hormone cycle. One way they disrupt this cycle is by reducing the level of testosterone, which lowers libido, depriving women of greater pleasure. A lower testosterone level also makes women less able to say "no" when they are asked to take on too much burden at work or at home. Especially concerning is that the longer a woman uses the Pill, the more her testosterone level falls.[2]

Another way oral contraceptives disrupt the natural hormone cycle is by depriving women of the estrogen spike that occurs during ovulation. This estrogen spike makes a woman look and feel more attractive. It also allows her to better discriminate among the scents of different men. This is important because a man's scent conveys valuable information about his major histocompatibility complex (MHC). Women are subconsciously attracted to men whose MHC profile is different from their own and complementary to it. The benefit of this complementarity is that any child they produce will inherit a more robust immune system. In the absence of an estrogen spike, a woman may end up marrying a man whose MHC profile is too much like her own. And if she subsequently goes off the Pill, she may suddenly find her partner's scent repellent.

Another negative feature of hormonal contraceptives is that they cause serious environmental pollution. The estrogen and progesterone in contraceptives are excreted in urine, which passes into sewage treatment plants and then into rivers and lakes where the chemicals disrupt the sensitive hormonal systems of fish, amphibians, and other aquatic life.[3]

Oral contraceptives do, however, have one significant benefit: they provide considerable protection against ovarian cancer. The risk of ovarian cancer is correlated with the number of times a woman ovulates. Using the Pill for 5 years will cut the risk of ovarian cancer in half.[4] It will also reduce the risk of uterine cancer. Prolonged nursing should also protect against ovarian and uterine cancer by suppressing ovulation.

The following table compares the copper IUD, the Pill, and lactation with respect to ovulation and menstruation:

	IUD	Pill	Lactation
Suppresses ovulation?	No	Yes	Yes
Suppresses menstruation?	No	No	Yes

Notes

1. The typical-use effectiveness rate of IUDs is 99 to 100 percent, oral contraceptives 93 percent, condoms 87 percent, diaphragms 83 percent, and the recently introduced Phexxi gel 86 percent. Reference: www.nytimes.com/2021/06/10/style/what-is-phexxi.html.

2. See pp. 80–82 in psychiatrist Julie Holland's book *Good Chemistry*.

3. According to the website www.sciencedirect.com/science/article/pii/S0160412016304494, "the world's human population discharges approximately 30,000 kg/yr. of natural steroidal estrogens and an additional 700 kg/yr. of synthetic estrogens solely from birth control pills. The amount of estrogen released into the environment from livestock is even higher than the amount released by humans. For example, in the United States and European Union, the annual estrogen discharge by livestock is 83,000 kg/year—more than twice the rate of human discharge." https://www.uofmhealth.org/health-library/tw9278.

4. See page 268 of *Ever since Adam and Eve* by physician Malcolm Potts and primatologist Roger Short.

Elimination of Tax Incentives and Disincentives for Reproduction

Many governments, including the United States, use the tax system to encourage reproduction. They offer tax reductions for such things as day care. Once a population credit system is in place, all tax measures that are intended to either promote or discourage reproduction would be eliminated. Couples will decide for themselves, free from government meddling, whether or not to have children. And poverty will be eliminated too, so everyone will have the opportunity to have children. (See the chapter "Eliminating Poverty.")

Prevention of Teenage Births

It still remains unrecognized, that to bring a child into existence without a fair prospect of being able, not only to provide food for its body, but instruction and training for its mind, is a moral crime, both against the unfortunate offspring and against society; and that if the parent does not fulfil this obligation, the State ought to see it fulfilled, at the charge, as far as possible, of the parent.

—John Stuart Mill[1]

Prior to the development of hormonal contraceptives and IUDs, the only way to securely prevent teenage births was for parents to closely monitor their teenage daughters and impress upon them the importance of chastity. Inevitably, these efforts sometimes failed, resulting in social disgrace for both the girl and her parents unless the pregnancy could be secretly terminated by an illegal abortion or hidden from view by sending the daughter away to a relative or a religious home for unwed mothers. The advent of effective contraceptives and the morning-after pill was profoundly liberating for both young women and their parents—indeed, for all fertile women.

That a woman should, at present, be almost driven from society for an offence which men commit nearly with impunity, seems undoubtedly to be a breach of natural justice. But the origin of the custom, as the most obvious and effectual method of preventing the frequent recurrence of a serious inconvenience to a community, appears to be natural, though not perhaps perfectly justifiable.

—Thomas Malthus[2]

The United States has the one of the highest rates of teenage births of any developed nation. In 2019, this rate was 18.8 teen births for every one thousand girls between 15 and 19. This represents a big improvement over the recent past: in 2011 the teen birth rate was 31.3 per thousand, while in 2008 it was 40.2, and in 1991 it was 61.8.[3] Currently, teen births represent about 10 percent of all US births. The industrial nations that are closest to the US in teen births are Ukraine (28 per 1,000), Russia (26 per 1,000), and New Zealand (26 per 1,000). Among the industrial nations with the lowest teen birth rates are Singapore, Slovenia, and Netherlands, all of which have 5 per 1,000; Japan and Hong Kong, which have 4 per 1,000; and Switzerland, which has only 2 per 1,000. At the other extreme, Burkina Faso has 128 teen births per 1,000, Kenya has 111, Malawi has 109, and Ethiopia has 91.[4] Child marriage is common in sub-Saharan Africa and South Asia, contributing significantly to high teen birth rates.

In the United States, children of teen mothers disproportionately end up in foster care. In Illinois, for example, 60 percent of foster care children were born to mothers below the age of 20, and 75 percent to mothers younger than 22.[5] The US teen birth rate for girls who have spent time in foster care is more than double the rate for those who have not. In Utah, this rate is three times higher.

One factor that contributes significantly to teen births is date rape. Another factor is a history of childhood sexual abuse. A 2009 meta-analysis of the relationship between childhood sexual abuse and teen pregnancy found that this abuse increased the average risk of teen pregnancy 2.21 times.[6] This has huge implications for countries like India and Pakistan, where both childhood sexual abuse and teen pregnancy are common. In a large study of child abuse conducted by India's Ministry of Women and Child Development,[7] it was found that 50 percent of India's children suffered emotional abuse, 69 percent suffered physical abuse, and 53 percent endured some form of sexual abuse. Fifty-seven percent of the victims of sexual abuse were males. About 20 percent of Indian childhood sexual abuse occurs in the home and 50 percent in institutional settings. Childhood sexual abuse in India is strongly correlated with poverty, which, in turn, is strongly correlated with overpopulation.

Another factor in high teen birth rates is membership in a conservative evangelical church. Conservative Christians strongly oppose premarital sex, so their young tend to marry early in order to enjoy sex or legitimize an out-of-wedlock pregnancy. These premature marriages undoubtedly contribute to the high divorce rate that prevails among evangelical Christians.[8]

Greatly reducing teenage births would generate a wide range of social benefits. It would help to expedite population reduction by increasing the spacing between generations. It would also reduce the number of foster children and the number of children who end up in poverty and prison.

Consider the murderer Charles Manson. He was born to an un-wed 16-year-old who had neither the mental nor material resources to care for him. As a child, he was passed from adult to adult and subjected to neglect and abuse. The result was a lifetime of crime and prison, culminating in the murders of seven people.

The most effective way to reduce the rate of teen pregnancy would be to ensure that all teen girls receive sex education and have ready access to the copper IUD and the morning-after pill.[9] These services should be dispensed by informed, empathetic adults that the teens can trust. Mentoring programs for at-risk teens would further help to reduce pregnancies.[10] All high school students should attend class-es in parenting skills. This should include plenty of practice with ro-botic babies that have to be cared for at home. Once teens learn how difficult and expensive parenthood is, they will be much less likely to get pregnant. And the acquisition of parenting skills will reduce child malnutrition and abuse, thereby helping prevent more children from becoming criminals. It is strange that our society requires no training at all for the socially critical task of parenting yet requires licensing for such relatively trivial enterprises as painting fingernails or cutting hair. Society eventually suffers from poor parenting, so parenting should not be regarded as a strictly private affair.

Something else we should do for teenagers is eliminate statutory rape laws. These laws have resulted in young men being placed on "sexual predator" lists that seriously damage their prospects in life. The very term "statutory rape" is dishonest because rape, by defini-tion, is forcible, whereas statutory rape is voluntary. Laws that ban voluntary sex should be replaced with a national law that requires those who have sex with a teenager to (a) use a condom and (b) sup-ply the girl (if it's a girl) with the morning-after pill. That would keep teens safe from both pregnancy and STDs. We need to acknowledge that teenagers have adult hormone levels and therefore adult sexual desire. Instead of stigmatizing and criminalizing their sexuality, we should make it safe and enjoyable.

The vicious cycle of teen births leading to foster children, leading to more teen births—all accompanied by great pain and immense cost to society—needs to be dealt with decisively. The cost of putting a halt to it would be far less than the economic and moral cost of letting it continue.

Notes

1.   John Stuart Mill, *On Liberty* (1858), 117.

2.   Thomas Robert Malthus, *An Essay on the Principle of Population* (1798), book 3, chapter 2.

3.   See https://www.guttmacher.org/news-release/2016/us-teen-pregnancy-birth-and-abortion-rates-reach-lowest-levels-almost-four-decades.

4.   The statistics on teen births in other countries are from a peer-reviewed paper titled "Adolescent Pregnancy, Birth, and Abortion Rates across Countries: Levels and Recent Trends," published under the auspices of the Guttmacher Institute, New York, and available to read at https://www.ncbi.nlm.nih.gov/pmc/articles/PMC4852976/.

5.   See https://www.chapinhall.org/research/children-of-young-parents-in-care-at-higher-risk-of-child-welfare-involvement/. Chapin Hall, University of Chicago research.

6.   Jennie G. Knoll, Chad E. Shenk, and Karen T. Putnam, "Childhood Sexual Abuse and Adolescent Pregnancy: A Meta-Analytic Update," *J Pediatr Psychol* 34, no. 4 (May 2009): 366–378. Published online

on September 15, 2008. Can be read at http://www.ncbi.nlm.nih.gov/pmc/articles/PMC2722133/.

7. L. Kacker, N. Mohsin, and A. Dixit, "*Study on Child Abuse: India 2007,*" Ministry of Women and Child Development, Government of India.

8. Michelle Goldberg, "Christian Marriage, Christian Divorce," *The Nation*, February 10, 2014, 6–8.

9. In the United States, the morning-after pill costs $40 or more. That is more money than many teen girls can afford. As a result, the morning-after pill has been a frequent target for shoplifting. In response, many drugstores now keep this product locked up behind the front counter.

10. David L. Kirp, "The Kids Are All Right," *The Nation*, February 28, 2011, 22–26.

Immigration under the Two-Credit System

Setting a Quota for Legal Immigration

Under the two-credit system, if there were no quota on legal immigration, and if all citizens were to donate or sell their credits to foreigners, a complete turnover of the population from natives to foreigners would take place once the natives died off. In reality, of course, most people would use their credits for reproduction, not donate them or sell them. In any case, I believe an annual quota should be established to limit the number of credits foreigners could purchase. We should keep track of how many people move out of the nation each year and limit immigration to that number.

Procedures for Legal Immigration

Before purchasing credits or receiving them as a donation, a prospective immigrant would register at the nearest US consulate. There, the immigrant's identity would be confirmed, a DNA sample would be taken, and a credit account would be set up for the immigrant. Once the account was set up, a US citizen could donate two credits to it, or the immigrant could purchase two credits to deposit in it.

Citizenship would be conferred as soon as the immigrant arrived at the port of entry and the credits were paid.

Upon becoming a new citizen, the immigrant, if less than 50 years old, would receive two population credits, just as a newborn baby would. The immigrant would then be able to use the credits in all the ways other citizens can.

If a foreigner who is older than 18 acquires two credits but then dies before he or she can immigrate to the US, the credits would be retired, just as those of a US citizen would. If the foreigner who received donated credits was under the age of 18 at the time of death, the credits would be returned to the donor, just as they would in the case of citizens.

Immigrants Who Bring Children

The situation of an immigrant who comes to the US accompanied by one child would be handled as follows: the immigrant, if under the age of 50, would receive two population credits when he or she received citizenship at the port of entry. The new citizen would then immediately use those two credits to pay for the child, allowing the child to also be granted citizenship. The child, however, in contrast to other immigrants under age 50, would receive no population credits. Once the child turned 18, if he or she wished to have children, the following options would be available: 1) use the two credits of a spouse or partner, 2) purchase two credits, 3) receive a donation of two credits, 4) adopt a baby or an embryo, or 5) return to his or her native country to have children there.

In the above scenario, both the immigrant and his or her child would become citizens without causing any long-term increase in population size. That's because the child would receive no population credits and the parent would no longer have any credits available for reproduction.

Immigrants who wish to come to this country with more than one child would first have to obtain credits for the extra children. This could be done by purchasing credits or receiving them as donations.

Elimination of Legal Residency Status ("Green Cards")

The category of *resident legal alien* would be eliminated. Therefore "green cards" would no longer be issued. Upon payment of two credits, a foreigner would become a US citizen at the port of entry. (There would, of course, be a security check first.). Alternatively, foreigners could enter the country for a limited time on a visa (tourist, student, etc.). Current holders of green cards would be given the opportunity to immediately become citizens as soon as the two-credit plan goes into effect. If they reject this opportunity, they would have to return to their home countries. If they subsequently wished to return to the United States, they would have to apply for a visa and, in most cases, would only be able to come as short-term visitors.

No More Filial "Anchors"

Under the two-credit system, the present practice whereby the children of illegal immigrants, upon reaching adulthood, sponsor their own parents or other relatives for legal immigrant status would cease.

Instead, all immigrants would have to be paid for with credits. There will be no more filial "anchoring."

Entry of Foreign Visitors

Tourists, scholars, scientists, craftsmen, artists, entertainers, and anyone else who has something to contribute, or is merely curious, would be welcome to visit, study, or (in a few cases) temporarily work in this country. However, a saliva sample for DNA record keeping would be taken from all visitors in order to identify them in case they illegally reproduce during their visit. Female visitors of reproductive age would also be screened for pregnancy. If pregnant, they would have to leave the US well before giving birth. If a female visitor gives birth while in this country, she will be expelled along with her baby and the father (if he too is a foreigner). The expelled baby would be allowed to return after the age of 18, because the Constitution currently confers citizenship on every baby born in this country. However, these returnees would not be issued population credits.

Even though visitors will be using the nation's resources during their stay (and therefore will function like additions to the population), it is probable that a comparable number of US citizens will be simultaneously visiting or living in other countries, thereby canceling any net detrimental impact.

Marriage to Foreigners

Currently, when an American wishes to bring a foreigner to the United States for marriage, the foreigner must apply for a K-1 visa. The foreigner is then interviewed at a US consulate to try to determine whether the intention to marry is sincere or merely a sub-terfuge for gaining permanent resident status. Once the K-1 visa is granted, the foreigner is allowed to come to the US for a period of 90 days. During that period, he or she must marry the American or return home.

Under a credit-based population-stabilization plan, the K-1 visa would be eliminated. The foreign partner/spouse would be treated exactly like any other immigrant. The American sponsor would need to deposit two population credits in the foreigner's account, and the foreigner would then be granted US citizenship upon arrival. Once the immigrant possessed citizenship, he or she would be at liberty to marry or not. This would eliminate the pressure to marry imposed by the present 90-day K-1 deadline. Some foreigners will undoubt-edly take advantage of their American sponsor by promising true love only to abandon the American as soon as they acquire citizen-ship, but at least this would save the American the trouble and ex-pense of a divorce.

The fact that a sponsor would have to pay two credits to bring a would-be spouse to the US does not mean that the new couple could not have children. If the foreign spouse is under the age of 50 on the date of arrival, he or she would be granted two credits, just like any other immigrant or newborn baby. The couple could use those credits to have one child (in effect allowing the American to replace himself or herself in the population, but not the foreigner). If they wanted to have more than one child, they would need to acquire additional credits or adopt a child or embryo for whom the credits had already been paid by the biological parents. The fact that they could have only one child (unless they acquired additional credits) might persuade more Americans to marry other Americans rather than foreigners.

Adoption of Foreign Children

In order to adopt a foreign child, the adopting couple (or individual) would first have the US consulate in the child's country set up an account for the child. The adopting parent(s) would then deposit two credits in the child's account. When the child entered the US, he or she would immediately be granted citizenship and would receive two population credits.

Refugees and Asylum Seekers

> Thinking hearts are better than bleeding.
> —Garrett Hardin[1]

US immigration law distinguishes between asylum seekers and refugees. The former are foreigners who have crossed into the US and claim they will be persecuted if they are sent home; the latter live outside the US and seek to come here because they are dispossessed or suffering persecution. The US refugee ceiling for 2022 was 125,000 refugees. The president determines the ceiling for the coming year and submits it to Congress for approval. Although the US has a ceiling on refugees, it has no ceiling on asylum seekers. In recent years, illegal immigrants have discovered that if they claim asylum status (saying they fear for their lives back home), they are processed by the Border Patrol and then typically released to await a hearing (a process that now takes about six years). In the meantime, many work illegally in the underground economy while their children attend American schools. Many of them will also breed, and their babies will receive automatic US citizenship. When these babies grow up, they will be able to sponsor their parents for legal residency even if the parents are eventually denied asylum and deported. And the Border Patrol? Instead of detaining and deporting illegal immigrants, their job has increasingly shifted to processing asylum seekers.

Under a credit-based population-stabilization program, the asylum program would undergo major reform. Asylum claimants, instead of being released into American society, would be held in camps until the validity of their claims could be determined. The process of researching their claims should take no more than a month. While staying in the camps, women would be segregated from men. Children would stay with their mothers (or fathers if there is no mother). Most asylum claims will likely be rejected as baseless, and the asylum seekers will be sent home. The word will then quickly spread in their home countries, and the flow of asylum seekers will slow to a trickle. The few asylum seekers with legitimate claims will be allowed to stay, and they will be offered US citizenship, but they will not be given population credits.

Some asylum seekers from El Salvador and Honduras legitimately fear gang violence. To address that concern, the US should invest in establishing honest, effective police forces and equitable judicial systems in Central American countries. The asylum seekers themselves might make good police recruits. Female asylum seekers should be offered free long-term contraception and sterilization. They should also be offered the opportunity to receive training as providers of reproductive services such as birth control, abortion, and vasectomies.

Then, when they are sent home, the US could pay them a salary based on the number of people they can persuade to adopt long-term contraception or get abortions.

As for refugees, they would first have to obtain two credits before they could come to the US. Because most refugees are poor, they would have to rely on credit donations or donations of money to buy credits. Most refugees are products of overpopulation, so the best way to stem the tide of refugees is to make long-term contraception and sterilization available at no cost in their home countries.

Tyrannies are notorious for producing refugees. They produce them by persecuting political opponents, provoking rebellions, and launching wars. It is tempting to think that this problem could be solved simply by killing the tyrant. Unfortunately, the problem of tyranny is deeper than that: it stems from the submissive psychology of the tyrant's people.

> Even when with great difficulties they have rid them-
> selves of the importunity of one master, they run to
> supplant him with another, with similar difficulties,
> because they cannot make up their minds to hate
> domination itself.
>
> —Michel de Montaigne[2]

When people flee their country, wealthy nations should respond by paying a neighboring country to house the refugees in camps until the problem that caused them to flee can be corrected. The administrators of these camps should never permit the occupants to breed—any more than we would permit animals in a shelter to breed. Yet, astonishingly, refugees in today's camps are given complete freedom to reproduce. In the world's largest refugee camp—Dadaab—in northern Kenya, Somali refugees are producing about 1,000 babies a month! These refugees say it is up to Allah to determine family size and that no matter how many children they produce, "Allah will provide." But, in fact, it is the wealthy nations that provide. Obviously, the women in these camps should be given mandatory long-term contraception (and mandatory abortions if they tamper with the contraception) or else the sexes should be segregated. And both sexes should receive mandatory reproductive and demographic education. The camp occupants should also be offered modest cash payments to persuade them to return home. It should be made clear to those who remain behind that they will never be released except to return home. Meanwhile they should be taught some skills that might be useful upon their return. For example, they could be taught how to provide contraceptive services. Then, when they return home, wealthy nations (or NGOs) could pay them a wage based on how many women they can persuade to adopt long-term contraception. Obviously, such a program would need to be carefully supervised to prevent fraud.

In the second (1806) edition of *An Essay on the Principle of Population*, Thomas Malthus described the moral dilemma that refugees pose for their hosts:

> A man who is born into a world already possessed, if he cannot get subsistence from his parents on whom he has a just demand, and if the society do not want his labour, has no claim of right to the smallest portion of food, and, in fact, has no business to be where he is. At nature's mighty feast, there is no vacant cov-

er for him. She tells him to be gone, and will quickly execute her own orders, if he does not work upon the compassion of some of her guests. If these guests get up and make room for him, other intruders immediately appear demanding the same favour. The report of a provision for all that come, fills the hall with numerous claimants. The order and harmony of the feast is disturbed, the plenty that before reigned is changed to scarcity; and the happiness of the guests is destroyed by the spectacle of misery and dependence in every part of the hall, and by the clamorous importunity of those who are justly enraged at not finding the provision which they had been taught to expect. The guests learn too late their error, in counter-acting those strict orders to all intruders issued by the great mistress of the feast, who, wishing that all guests should have plenty, and knowing she could not provide for unlimited numbers, humanely refused to admit fresh comers when her table was already full.[3]

In 2015 and 2016, more than three million refugees and economic migrants from Syria, Afghanistan, and North Africa flooded into Europe. Most of them headed for Germany and Sweden because those countries "have gotten up to make room" for the refugees. As Malthus recognized, this only encouraged even more immigration and soon brought an end to the happiness of the hosts. The two-credit system would prevent the very existence of the unwelcome guests by preventing them from being born.

If all our efforts to prevent the creation of refugees fail, and no nation can be found to shelter refugees even for money, a wealthy nation could hold a referendum to determine whether to admit more refugees. If the vote favored admitting them, the government would purchase the necessary number of credits and give them to the refugees. This credit purchase would reduce the domestic supply of credits, resulting in a lower birth rate and fewer nonrefugee immigrants. Thus, the influx of refugees would be exactly compensated by a decrease in births and a decrease in conventional immigration. Purchasing credits for refugees would be a substantial public expense, but the money would remain inside the nation as it was transferred from the taxpayers (represented by the government) to the willing sellers of credits. But we must never forget that it is parents who are ultimately responsible for their children (and upon whom the children have "a just demand"). Those who cannot afford children should not be having them.

Notes

1. Garrett Hardin, *Naked Emperors: Essays of a Taboo-Stalker* (Los Altos, CA: William Kaufmann, Inc., 1982), 29.

2. Michel de Montaigne, *Michel de Montaigne: The Complete Works*, trans. Donald M. Frame (New York: Alfred A. Knopf, 1943), 101. (The quotation is from *The Essays*.) Tyrannical governments arise in cultures where the structure of the family is also authoritarian. In other words, families where a paterfamilias controls his wife and children through threats of violence. Throughout the world, the most common and accepted form of domestic violence is spanking—an ancient practice whose antecedents can be found in chimpanzees. When a male chimpanzee assaults a female, the female tries to convert his dangerous physical aggression into safer sexual

aggression by presenting her buttocks for sex. Natural selection has favored this trick and reinforced it by making it erotically exciting. The blows of spanking mimic the rhythmic thumping of the buttocks that occurs during sex. Spanking combines the excitement of sex with the excitement of violence. This link between sex and violence helps explain why so many women (and their partners) are turned on by spanking and why so many women—to their shock and chagrin—experience orgasms when raped. It also helps explain why abducted women may fall in love with their captors (the Stockholm syndrome). Throughout the world, humans express their submission to a greater power by "assuming the position." Hence, we have the Chinese kowtow, the Japanese *zarei*, and the various "butts up" postures adopted by the faithful when they pray.

3. Thomas Robert Malthus, *An Essay on the Principle of Population*, 2nd ed. (1806), book 4, chapter 6.

Downsizing the Immigration Bureaucracy

The two-credit plan would supersede nearly all existing immigration laws and policies, and radically shrink the immigration bureaucracy and the tribe of immigration lawyers. Only a small immigration department would remain to handle visas for temporary residence. There would also be a small corps of consular officials who would attend to setting up accounts for foreigners desiring to immigrate.

Summary of the Government's Implementation Responsibilities

If citizens, acting through their elected representatives, approve a population reduction plan like the one proposed in this book, the duties of the government would be these:

- Carry out registration and credit distribution
- Operate the online credit market
- Keep the vital statistics database up to date as babies are born, immigrants arrive or leave, and people die
- Facilitate online and telephone inquiries about population-credit status
- Process and record DNA samples
- Promote reproductive education
- Oversee sperm bank contracts
- Sterilize violators
- Prevent illegal immigration (primarily by enforcing laws against employing illegal immigrants)
- Make effective population control a precondition for foreign aid
- Purchase and then retire credits in order to eliminate population momentum during the early years of the plan's implementation

- Regulate visits by foreigners
- Provide an abortion tax checkoff, so people who have ethical objections to abortion will not have to pay for it
- Provide full contraceptive, sterilization, and abortion services at every hospital that has a maternity ward.

The government already has two of these responsibilities: to secure our borders against illegal immigration and to regulate visits by foreigners. The other responsibilities would be new.

The government's role regarding the two-credit system would be that of an umpire: keeping the game fair by enforcing the rules. The players would be individual citizens, utilizing their reproductive credits in whichever way they choose. The public could periodically vote on the rate at which they want their population to shrink and the number of immigrants they want to admit annually.

The government staffing required to implement a population-stabilization plan would be quite small. Computers would carry out most of the data handling, which would reduce to a minimum the opportunities for human error and mischief. Most of the human staff would consist of registrars assigned to every maternity ward. In many cases, these registrars would be medical doctors who already work at the hospital and who have taken special training to qualify as registrars.

The overall costs of operating the two-credit system would be small and would be recovered quickly by the social and environmental savings brought about by reducing population size. These savings would increase with time as the population steadily became smaller.

How the Two-Credit System Can Cut the US Population in Half in 70 Years

> You say you got a real solution. Well, you know, we'd all love to see the plan.
>
> —The Beatles, "Revolution"

The population of the United States is currently growing at about 0.5 percent annually. The two-credit system can halt this growth within one year and then begin shrinking the population. If we decide on a reduction rate of 1 percent per year, the population will shrink to half its present size in just 70 years.

Starting with the 2021 US population of 332 million, we would need to reduce the number of births and new immigrants by 3.3 million during the first year in order to meet a 1 percent reduction target. Multiplying 3.3 million by 2 gives us the number of credits—6.6 million—we would have to retire the first year. This credit reduction would come from two sources: 1) government purchases of credits from willing sellers and (2) credit attrition. The ratio of government purchases to attrition would be about 8 to 1 throughout most of the period of population reduction. However, once the population approached its optimal size, government purchases would no longer be necessary, because credit attrition alone would suffice for further population reduction.

Credit attrition would occur by several mechanisms. One of them stems from the fact that each person would be issued exactly 2.0 population credits even though the replacement fertility rate in modern societies is 2.1 children per woman. The discrepancy between 2.1 and 2.0 means that for every baby born under the plan, 5 percent of a baby would not be born that would have been born if the population were exactly replacing itself (0.1/2.0 = 0.05). Given that 3,659,289 births were recorded in the US in 2021, that would mean that 182,964 (0.05 × 3,659,28940) fewer babies would be born than if the 2.1 fertility rate were in play. That's nearly a fifth of a million fewer babies annually due solely to the small 0.1 deviation from true replacement level.

A second mechanism of credit attrition would be the deaths of people between the ages of 18 and 50 who have not yet used both of their credits. In such cases, the unspent one or two credits of the deceased would automatically be retired, making the population reduction represented by their deaths permanent. Roughly 71,000 Americans under the age of 25 die every year.[1] Most of them would not have used their credits, so we can safely assume that about 140,000 credits would be permanently retired each year due to these premature deaths. Of course, some of these unfortunate young people would already have used their credits, but this would be counterbalanced by the many who die prematurely at older ages (between 25 and 50) with unused credits in their accounts which would be retired. So 140,000 retired credits is a conservative figure.

A third source of credit attrition would be people who turn 50 without having used both their credits. Their unused credit or credits would be permanently retired. This group would include altruists who have decided not to use their credits in order to help speed population reduction and people with severe mental impairments who were incapable of using their credits. We cannot know how many altruists there would be (although the number would likely be small), but we do know that about 3 percent of Americans are mentally retarded and probably not fit to use their credits.

The table on the next page shows the total number of credits that would be retired by attrition, and the number of credits the government would have to purchase over the course of 350 years (2021-2371).

Table Notes

Note 1: When we add up all the deaths of children 0 to 14 and of teens and young adults 15 to 24, the total per year is about 70,826 (rounded to 71,000 in the table). See www.worldlifeexpectancy.com/usa-cause-of-death-by-age-and-gender.

Note 2: In 2021, 3,659,289 babies were born in the US (https://www.cdc.gov/nchs/nvss/births.htm). Multiplying this by 0.05 gives us 182,964 (rounded to 183,000 in the table). This is the number of babies born in 2021 who would not have been born due to the discrepancy between the 2.0 credits that people would receive, and the 2.1 credits needed for replacement-level fertility in industrialized countries.

Note 3: About 3 percent of the US population is classified as mentally retarded (https://mn.gov/mnddc/parallels2/pdf/80s/82/82-PMR-ARC.pdf). These are people whose IQ is less than 70. In 2021, a total of 3,458,697 Americans died. Three percent of these, or 103,761 would have been mentally retarded (rounded to 104,000 in the table).

Note 4: This is just a hopeful guess.

		Year (in 70-year intervals)					
		2021	2091	2161	2231	2301	2371
	US population in millions (halving every 70 years)	332	166	83	41.5	20.75	10.375
Credit Attrition	Annual deaths of Americans 0-25 years old. See note 1.	71,000	35,500	17,750	8,875	4,437.5	2,218.75
	People not born annually due to the discrepancy between 2.0 credits and 2.1 replacement-level fertility. See note 2.	183,000	91,500	45,750	22,875	11,438	5,718.75
	Annual deaths of the mentally impaired whose unused credits would be retired. See note 3.	104,000	52,000	26,000	13,000	6,500	3,250
	Annual deaths of non-reproducing altruists who 2 unused credits would be retired. See note 4.	20,000	10,000	5,000	2,500	1,250	625
	Annual no. of people whose existence would be prevented by credit attrition (sum of the preceding 4 rows).	378,000	189,000	94,500	47,250	23,625.5	11,812.5
	Annual no. of credits lost to attrition (equals preceding row × 2).	756,000	378,000	189,000	94,500	47,251	23,625
	No. of credits (in millions) that would need to be retired to reduce the population at 1 percent annually (equals 1 percent of the population size × 2)	6.64	3.32	1.66	0.83	0.415	0.2075
	No. of credits (in millions) that the government would need to purchase annually (subtract from annual no. of credits lost to attrition)	5.884	2.942	1.471	0.7355	0.367749	0.183875

The above table indicates that if we initiated the two-credit system in 2021, 0.756 million credits would be retired by attrition the first year. But that number is only 13 percent of the 5.792 million credits that we need to retire the first year in order to reduce the population by 1 percent. The remaining 87 percent (5.039 million credits) would have to be purchased by the government from willing sellers, and then retired. In subsequent years, the government would need to purchase far fewer credits, because the population would be shrinking. This is shown in the bottom row of the above table.

When we graph the projected population decrease of the US at the 1 percent rate, it shows a classic exponential growth curve—but in reverse:

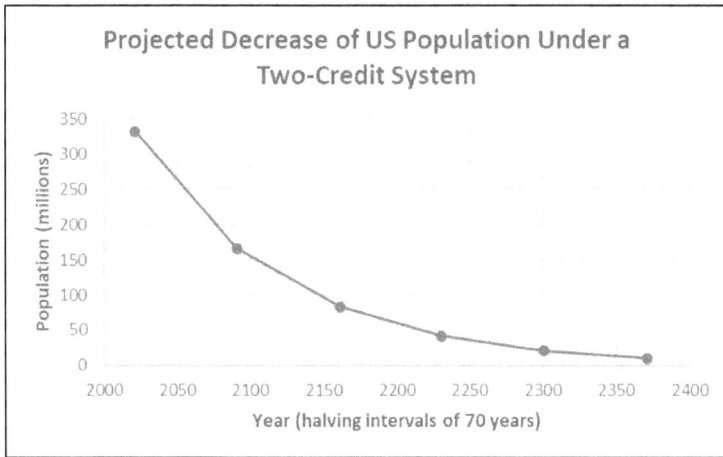

The chart shows that there would be large population reductions in the early years, which would gradually taper off. The first 70 years would see the US population cut by 166 million, while the last 70 years (years 2301 to 2371) would see it cut by only 10.4 million.

The annual cost to the taxpayers for credit retirement would depend on the market price of credits. The following table gives an idea of what these costs might be for the first year, using hypothetical credit prices:

Cost per credit	Cost per credit × 5.792 million credits
$10,000	$57.9 × 10^9$
$20,000	$115.8 × 10^9$
$30,000	$173.7 × 10^9$
$40,000	$231.7 × 10^9$
$50,000	$289.6 × 10^9$
$100,000	$579.2 × 10^9$

Government purchases of credits would represent a large expenditure, but this expenditure pales in comparison with the cost of failing to rapidly shrink our population. By comparison, the US spends nearly $800 billion on its military each year (a good deal of which is poorly spent).

The next question is whether a sufficient number of Americans would be willing to sell their credits for the government to be able to purchase and retire 5.792 million credits the first year. To answer that, we need to know how many Americans never produce children or produce only one child, because nearly all of them would opt to sell their remaining credit(s) before they turn 50 (although a small percentage will choose to donate them or do nothing with them). Demographic data show that as of May 2015, 15 percent of American women aged 40 to 44 never produced a child,[2] and 18 percent produced only one child.[3] The percentages for male procreation would naturally be quite similar. Fifteen percent of the 2021 US population of 332 million is 49.8 million. So there would be 49.8 million people who would have two credits available to sell. That would be 99.6 million credits. And the 18 percent who produce only one child represents 59.8 million people who would have one credit to sell. Adding these two sources of credits together would give us 99.6 + 59.8 = 159.4 million credits that would likely be offered for sale by these two groups of people sometime between the ages of 18 and 50 (a span of 32 years). If we divide these 159.4 million credits by 32 years, we get the average number of credits that would be offered for sale each year. The answer is 4.98 million credits. That figure is shy of the 5.792 million credits the government would need to purchase during the first year. However, it would be easy to obtain the remaining 0.81 million credits, because the introduction of universal reproductive education and the universal availability of free contraception and abortion would greatly reduce the current 18 percent rate of unwanted (as opposed to mistimed) pregnancies.[4] This would allow millions of additional credits to become available for sale, which would enable a faster rate of population reduction even beyond 1 percent annually should we choose to do so. Moreover, the government could always obtain and retire more credits simply by offering a higher price for them.

Malthus believed that only hunger could bring about long-term population decline: "There has never been, nor probably ever will be, any other cause than want of food which makes the population of a country permanently decline."[5] But we have now seen that the two-credit system can bring about long-term population decline without hunger. Indeed, the eradication of hunger would accompany population decline.

NOTES

1. For data on the death rates of young people see www. worldlifeexpectancy.com/usa-cause-of-death-by-age-and-gender.

2. See https://www.pewresearch.org/social-trends/2015/05/07/ childlessness-falls-family-size-grows-among-highly-educated-women/. This article was published by Pew Research Center on May 7, 2015. A little over a year earlier, on January 3, 2014, Pew published an article that put the US childless percentage at 19 percent, rather than 15 percent.

3. Ibid.

4. In 2011, 45 percent of the 6.1 million pregnancies in the United States were unintended. See "Declines in Unintended Pregnancy in the United States, 2008–2011" by Lawrence B. Finer, PhD, and Mia R. Zolna, MPH, *N Engl J Med* 374, no. 9: 843–852, March 3, 2016 (https://www.nejm.org/doi/full/10.1056/NEJMsa1506575).

5. Donald Winch, ed., *An Essay on the Principle of Population* (Cambridge University Press, 1992), 195.

Part 3:

Demographic and Genetic Consequences of the Plan

Variables Controlling the Rate of Population Decrease

Once a credit-based population reduction plan is put into operation, four variables will control the rate at which the population decreases: (a) the mortality rate, (b) the credit-expiration rate, (c) the birth rate, and (d) the net migration rate. The higher the mortality rate and credit-expiration rate, and the lower the birth rate and net migration rate, the faster the population size will decrease. This can be expressed in mathematical form:

rate of population decrease = [mortality rate + credit-expiration rate] − [birth rate + net migration rate].

The net migration rate is simply the difference between the number of people moving into the country (immigration) and the number exiting the country (emigration). Immigration is demographically comparable to birth, and credit expiration to death. Credit expiration prevents the replacement of an individual in the population, thereby reducing the size of the population. Credit expiration will be a new concept for demographers because they have traditionally held that populations can only decrease by a higher death rate, a lower birth rate, or a lower net migration rate.

The credit-expiration rate could be raised or lowered by the government in response to a vote of the people. If the people voted to increase the credit-expiration rate, the government would purchase more credits and retire them.

For populations that are declining at a steady rate, the same mathematical formula used to compute doubling time can be used to compute halving time. For example, if the 2021 US population of 330 million were declining at 0.5 percent per year (which is the rate at which it was increasing in 2021), it would reach the half point (165 million) in 139 years (69.3 ÷ 0.5). (Recall that 69.3 is 100 times the natural logarithm of 2.) The last time the US had a population of 165 million was in 1955.

With a two-credit system in place, the population would continue to shrink until it eventually reached the optimal size—the size at which the population could live compatibly with all other species and still defend itself and enjoy a rich cultural life. This size will depend on the population's level of resource consumption and pollution: the higher the consumption and pollution, the smaller the optimal population size must be. Once an optimal population size has been attained, the number of credits annually retired would be reduced in order to halt further population decrease and keep the population stable. This could be accomplished by recycling some retired credits through a public drawing. The winners could have additional children or donate or sell the credits. On the other hand, if the mortality rate began to fall (perhaps because people adopted a healthier diet or there was a major advance in medical treatment), the government could purchase more credits and retire them in order to keep the population from growing beyond the optimal size. And if the mortality rate were to rise (perhaps because of a lethal epidemic), retired credits could be put back into circulation. These adjustments in credit supply could be made automatically by computers linked to a continually updated vital statistics database.

For a population at equilibrium, the number of people annually added to the population through births and immigration must always equal the number removed through deaths and emigration. When a population is at equilibrium, and the rate of emigration is held equal to the rate of immigration, the rates of birth and death will be identical and equal to the reciprocal of life span × 1,000. For example, here in the United States, where the average life span is currently 79 years, the equilibrium rate of birth and death would be 1/79 × 1,000 or 12.7. So 12.7 births and deaths per thousand people per year would indefinitely maintain the US population at its present level (assuming no net migration). For comparison, the actual US birth rate in 2022 was 12.3 per thousand, and the death rate was 8.4 per thousand, so the population is still growing rapidly.[1] Immigration is an even bigger factor than domestic births in driving US population growth.

Notes

1. CIA World Factbook.

Elimination of Population Momentum

When a rapidly growing population abruptly switches to replacement-level fertility (about 2.1 children per couple in industrialized countries), the population will continue to grow for about 70 years (one lifetime). This time-lag phenomenon is called population momentum. It is due to the high proportion of young people in populations that have undergone rapid growth. In other words, even though each woman is having fewer babies, there are more women having babies. Population momentum dissipates gradually as the people who were alive at the time of the growth spurt die off. Once they are gone, the new rate takes over.

Population momentum is not inevitable: it can be slowed quite rapidly (and even reversed) by lowering the fertility rate to well below the replacement level and limiting immigration. Even when fertility remains at the replacement level, population growth can still be substantially lowered by increasing the age at which people marry and, to a lesser degree by increasing the spacing between births.[1]

Population momentum is analogous to the slow clearing of a traffic jam once the number of vehicles entering the highway becomes less than the number exiting.

A credit-based population-stabilization plan would counteract population momentum in more than one way. First, it would automatically keep the birth rate below the replacement level of 2.1 by limiting births to an average of 2.0 per woman (since people would have 2.0 credits). Second, it would retire the credits of those who don't use them by the age of 50, as well as those who die between the ages of 18 and 50. Third, the plan combats population momentum by reducing births by the youngest women, thanks to universal sex education and universal availability of contraception and abortion. Fourth, the plan encourages large spacing between the second and third child, because parents who want a third child would likely need to save for a number of years in order to purchase the necessary credits. Fifth, the plan requires that immigration occur at the expense of births, not in addition to births. In other words, more immigration would mean correspondingly fewer births. All of these factors combined would substantially reduce population momentum. And if they proved insufficient to completely eliminate population momentum, the government could purchase and retire the necessary number of credits to eliminate it. These purchases would be made from willing sellers in the credit market. If such government purchases proved necessary, the effect would be to reduce the overall supply of credits, thereby boosting the price. That would tempt more people to sell their credits, which would bring the credit price back down.

One way to make government purchases of credits more democratic would be for people to periodically vote on whether they want their population to remain at its current level or undergo further decline at a specified rate. The ballot could offer several rates of population reduction for voters to choose from.

Notes

1. Joel E. Cohen, *How Many People Can the Earth Support?* (New York: W. W. Norton & Company, 1995), 142.

Immigration During the Period of Population Reduction

Both during and after the period of population reduction, immigration would continue. The immigration rate would depend on whatever annual quota we set for immigrants. In contrast to the situation today, immigrants arriving under the two-credit plan would mostly be well-educated people of substantial means (those able to afford credits). This would be a significant benefit for the receiving country (and a loss for the home countries of the immigrants).

Any increase in the domestic birth rate would result in a decrease in the immigration rate, because fewer credits would be available for sale or donation to immigrants.

Impact of the Two-Credit System on Population Genetics

> Obscure as is the problem of the advance of civilization, we can at least see that a nation which produced during a lengthened period the greatest number of highly intellectual, energetic, brave, patriotic, and benevolent men, would generally prevail over less favored nations.
>
> —Charles Darwin[1]

The more children people produce and successfully raise to adulthood, the greater their contribution to the overall genetic makeup of the population (including average intelligence level). Under a two-credit system, the genetic contribution of those who are most inclined to breed would be lessened, because they could not have more than two children unless they acquired additional credits. But even so, they would still likely produce their first two children in their early twenties, which might give them sufficient time to earn enough money to buy credits to have more children. And by the time they produced their last child, their first child would already be old enough to reproduce. So those who are most eager to reproduce will still dominate the gene pool, but not as much as they presently do, and no longer at society's expense.

Impulsiveness is a trait that every society would benefit from reducing. It contributes to gambling, rape, traffic collisions, shoplifting, rudeness, addictions, and fighting. There is good evidence that impulsiveness has a genetic basis. If a two-credit system were in place, impulsive people would likely sell their population credits right away in order to enjoy spending the money. Having sold their credits, they would not be able to produce offspring who would inherit their impulsiveness. This would increase overall social well-being and allow reductions in insurance rates (medical and automobile) and reduce the costs of the criminal justice system (police, courts, prisons).

Another way the two-credit system would improve the gene pool is by reducing the average level of testosterone in society. It would accomplish this in two ways: First, it would automatically limit the number of children a man can produce to two (unless he acquires additional credits). This would prevent men with high testosterone levels from siring numerous offspring and thereby increasing the average testosterone level in the population. Second, the two-credit system would lower the average testosterone level by reducing the proportion of young people in the population. (A population that is shrinking always has a smaller proportion of young people than one that is growing.) Among the benefits of reducing the mean testosterone level of a society would be more responsible fatherhood, less violence against wives and children, less rape, less pedophilia, fewer incidents of road rage, lower rates of crime (especially assault and murder), and a lower likelihood of war. This would contribute to a more civilized and tolerant society.[2]

Another way the two-credit plan would improve a population's genetic status is by offering free genetic screening to all who register to obtain credits. This would enable couples to learn the likelihood that they would produce a defective child. Gene therapy may someday be able to correct genetic defects, but in the meantime, we should screen the DNA of all registered couples for the most serious genetic diseases (those that are certain to cause a painful and short life) and then require that all couples who carry these genes refrain from reproduction. Such couples should be compensated by being given priority for adoptions.

Obviously, no one wants alcoholics and drug addicts producing children. Not only do addicts make incompetent parents, but their children often have serious psychological and physical problems (such as fetal alcohol spectrum disorders). Moreover, the children of addicts are more prone to become addicts themselves, at least in part because there is good evidence that addiction has a genetic component. All female addicts should be required to use long-term contraception (such as an IUD or hormonal implant or sterilization). To secure compliance, the government should offer female addicts free drugs and free addiction treatment services if they agree to use long-term contraception. Preventing addicts from breeding would provide the following benefits: an addicted mother would not be burdened with a sickly child she could not properly care for; society would have lower expenditures for health care, foster care, criminal justice, and welfare; and, best of all, the genes that contribute to a predisposition to addiction would gradually decline in the population.

Now that the use of contraceptives has become widespread, natural selection has begun to favor the genes of those who reject contraception. Unfortunately, such people lack a well-developed social and environmental conscience. The two-credit system would help reverse this harmful trend.

All of us would like to increase the percentage of healthy, good-looking, intelligent, wise, responsible, friendly, and considerate people in our society. These are the people we would like to have as parents, friends, neighbors, and spouses. But many of these admirable people have chosen not to reproduce. The two-credit system would counter this by allowing them to contribute their genes to another generation without having to raise the children themselves. They would do this by contributing their sperm and egg cells for in vitro fertilization. The fertilized egg would be implanted in a woman who is infertile or who desires to have more than two children but lacks the credits to pay for them. The donors of the sperm and egg cells would pay the two credits, not the implanted woman. The latter and her spouse or partner (if she has one) would then raise the child as their own. This process is called embryo donation or embryo adoption. It is a form of adoption that begins right after conception rather than postbirth. Another difference from conventional adoption is that the recipient of the fertilized egg and her spouse (if she has one) would decide which features they would like the baby to have by researching a donor database. Sperm bank users already do this. We can be confident that the adopting parents would choose to have children with good looks, high intelligence, agreeable dispositions, and excellent physical health. Everyone would benefit from this: the donating couple would have the satisfaction of knowing that they had contributed desirable genes to society; the adoptive parents would be blessed with a child with excellent genes (and therefore excellent prospects); the child would have the double benefit of good genes and loving adoptive parents; and society would benefit from having a higher percentage of smart, healthy, wise, brave, kind people. In this way, the two-credit system would provide an effective, noncoercive way to improve the human genome. It would avoid the two historical problems with eugenics: coercion and racism. One caveat about this process of embryo adoption is that any woman who wishes to adopt an embryo would first have to pass a means test to demonstrate that she can afford to properly care for the child. Otherwise, there would be women who would serially adopt embryos without the wherewithal to support them.

Notes

1. Charles Darwin, *The Descent of Man*, chapter 5.

2. High testosterone levels do not inherently make people more aggressive. Rather they lower the threshold for aggression in those who are already predisposed to aggression. Specifically, testosterone reinforces whatever behaviors are required to maintain status in a particular cultural context. If aggression is required to maintain status, then testosterone lowers the threshold for aggression. But if status can be maintained without aggression, testosterone may instead turn men toward peaceful means of maintaining status. Another important fact about testosterone is that high levels make people overconfident and narcissistic, boosting impulsivity and risk-taking. For more on the role of testosterone, see chapter 4 of Robert Sapolsky's book *Behave: The Biology of Humans at Our Best and Worst*.

Part 4:

As the Population Shrinks

Moving toward a Stable (Steady State) Economy

> The endless-growth economy, contrary to orthodox belief, is a diseased economy.
>
> —Edward Abbey[1]

When a population shrinks, its economy will shrink to the same degree due to less demand for goods and services. The title of this book could just as well have been *Reversing Economic Growth Swiftly and Painlessly: A Simple Two-Credit System to Regulate Resource Consumption and Pollution.* As the population shrinks, material throughput (meaning consumption and pollution) will shrink correspondingly until a democratically determined optimal population size is attained. At that point, throughput will stabilize, and the economy will become steady state.

In a steady state economy, development does not cease but becomes qualitative rather than quantitative. Qualitative development consists of everything that improves well-being, such as advances in disease prevention and medical care, a more equitable distribution of wealth, better crime prevention, the flourishing of art and science, and new technologies that permit commodities to be manufactured with less material, less energy, less pollution, less labor, and more durability. Unlike quantitative development, qualitative development can continue indefinitely, because nature imposes no limits on it.

Among the many benefits of a shrinking population and economy are lower housing costs (for purchase or rent), more resources for ourselves and our descendants, lower public expenditures for education (thanks to fewer children), higher levels of employment (due to reduced labor competition and fewer hours worked per person), lower land prices (which would enable many more people to become independent farmers), cleaner air and water (hence better health for ourselves and other species), a stabilized climate, and reduced traffic congestion (resulting in better vehicle and worker efficiency, fewer collisions and injuries, and lower insurance costs). With fewer children, people would be able to save more for health care and retirement. Alternatively, they could afford to pay higher taxes so that the government could efficiently provide these services. Another major benefit of ending growth would be less crime, thanks to less unemployment and fewer young people.

Falling housing prices caused by a shrinking population would nearly eliminate the home-building industry. Home builders would instead be employed in deconstruction. They would disassemble houses and commercial buildings so that the raw materials could be stored for future domestic use or export. No longer would abandoned buildings be allowed to decay until they are eventually burned by vandals (as happened on a massive scale in Detroit and other US cities).

Among the materials that could be salvaged from abandoned homes, office buildings, and factories are machinery, appliances, lumber, bricks, cinder blocks, concrete, asphalt, metals, glass, furniture, and household goods. Our sales of salvaged lumber would largely eliminate the world's timber industry. This would allow the world's remaining forests to continue to protect biodiversity and serve as carbon sinks to counter climate change. Our sales of salvaged metals would put the world's strip mines and polluting smelters out of production. Our sales of salvaged concrete and asphalt would reduce sand and gravel mining and the consumption of crude oil used to make bitumen.

Former construction workers, besides being employed in deconstruction and recycling, could be employed in retrofitting existing homes with solar panels and solar water heaters, thereby reducing CO_2 emissions.

Most economists reflexively reject the idea that long-term economic contraction could be a good thing. They were trained to believe in the cornucopian fantasy that economies can grow forever (ignoring the fact that our world is finite). Their foolish faith in endless growth is why so many of them embrace economic bubbles and are always shocked when the bubbles burst.

> Our most serious problem, perhaps, is that we have become a nation of fantasists. We believe, apparently, in the infinite availability of finite resources.
> —Wendell Berry[2]

As a population shrinks, its labor force will decrease correspondingly. In the unlikely event that we need additional laborers, there are several ways we could obtain them without increasing population size. First, we could employ more of our young people, many of whom, for lack of jobs, are now in prison or marking time in college. Second, we could make retirement a gradual process, extending it over a decade or more. Third, the productivity of our existing labor force could be improved by greater use of computers and robots. Fourth, we could obtain more labor by restricting our labor to socially beneficial endeavors. For example, we could entirely eliminate the lawn-care industry by converting our lawns into organic vegetable gardens, orchards, small prairies, woodlots, or xeriscapes. This would free up workers in the lawn-care industry for more useful enterprises and would simultaneously reduce air pollution, water pollution (from fertilizer and pesticide runoff), noise pollution (from leaf blowers and lawn mowers), and conserve water, fuel, and fertilizer. It would also provide more habitat for animals. One thing we should never do to obtain more labor is admit guest workers. Every country that has done so has ruefully discovered that the guests soon make themselves at home and begin reproducing. Of course, this problem could be avoided if we admitted only gay males as guest workers.

In Japan, population reduction is already well underway. Since 2010, the Japanese population has been shrinking by one-quarter million each year. This decline is the result of a low birth rate coupled with an extremely low immigration rate. One of the many benefits of this decline is that housing costs are falling in Japanese cities. This is enabling many Japanese to move into the desirable city centers. Another benefit is low unemployment and high wages. The Japanese are showing the rest of the world that the end of population growth and economic growth is a good thing. Unfortunately, the Japanese economy is still far too dependent on foreign trade and the Japanese national debt has been allowed to grow too large.

There are certain to be many unexpected challenges as the world's economies contract, but with creativity and goodwill we can successfully cope with all of them. Once our basic physical needs are satisfied, what we really want and need are security, friendship, love, and a chance to develop artistically, intellectually, and spiritually. With goodwill, we can create a society that will serve *our* needs instead of serving the limitless greed of the rich.

> A sustainable society would be interested in qualitative development, not physical expansion. It would use material growth as a considered tool, not a perpetual mandate. Neither for nor against growth, it would begin to discriminate among kinds of growth and purposes for growth. It could even entertain rationally the idea of purposeful negative growth, to undo excess, to get below limits, to cease doing things that, in a full accounting of natural and social costs, actually cost more than they are worth.
>
> —Donella and Dennis Meadows
> and Jørgen Randers[3]

Notes

1. Edward Abbey, "The Conscience of the Conqueror," in *Abbey's Road* (New York: Penguin Books, 1979), 136.

2. Wendell Berry, *What Are People For?* (San Francisco: North Point Press, 1990), 202.

3. Donella Meadows et al., *Limits of Growth: The 30-Year Update* (White River Junction, VT: Chelsea Green Publishing, 2004), 255. Economist Thomas Piketty has called attention to the impact of low population growth and low economic growth on capital accumulation: "Conversely, a stagnant, or worse, decreasing population, increases the influence of capital accumulated in previous generations. The same is true of economic stagnation. With low growth, moreover, it is fairly plausible that the rate of return on

capital will be substantially higher than the growth rate" *(Capital in the 21st Century*, p. 84). In other words, without a universal basic income or some other mechanism to equitably distribute wealth, low population growth and low economic growth would allow the rich to continue getting richer. An economy that is not growing also decreases social mobility, because few new professions are created that would provide opportunities for upward mobility.

A Shift to Smaller Communities

Let there be small countries with few people.
—*Tao Te Ching*[1]

As populations shrink, people will once again organize themselves into smaller, friendlier communities. Edward Abbey imagined them as "small islands of civilization in a sea of unspoiled nature."[2] Aldous Huxley, too, dreamed of smaller communities:

> Over-population and over-organization have produced the modern metropolis, in which a fully human life of multiple personal relationships has become almost impossible. Therefore, if you wish to avoid the spiritual impoverishment of individuals and whole societies, leave the metropolis and revive the small country community, or alternatively humanize the metropolis by creating within its network of mechanical organization the urban equivalents of small country communities, in which individuals can meet and cooperate as complete persons, not as mere embodiments of specialized functions.[3]

Small communities are inherently more democratic than large ones, because everyone's voice can be heard. And when people in small communities have conflicts with their neighbors, they rarely cause much damage (unlike the devastating wars of large nations).

> The citizen of a small state is not *by nature* either better or wiser than his counterpart in a large power. He, too, is a man full of imperfections, ambitions, and social vices. But he lacks the power with which he could gratify them in a dangerous manner, since even the most powerful organization from which he could derive his strength—the state—is permanently reduced to relative ineffectualness."
>
> —Leopold Kohr[4]

Another advantage of small communities is that some epidemic diseases cannot survive in them. Measles, for example, requires an urban population of at least 500,000 to perpetuate itself.[5] In small states, sexually transmitted infections (STIs) could easily be eradicated. If this were done in combination with easy access to effective contraception, all members of the community could enjoy sex without fear of contracting a debilitating disease or becoming unintentionally pregnant. This would contribute to a much more relaxed and friendly society in which men and women could confidently share sexual pleasure.

Even when living in small communities, people will still need a democratic government to perform three basic functions:

- Prevent population increase

- Prevent a rogue community from trying to dominate other communities
- Prevent environmental degradation.

Such a government would function much like an immune system, detecting instabilities in the system and moving quickly to correct them. This is as close as mankind can ever hope to come to the ideal of maximizing personal freedom and responsibility. Some societies, such as that of the Pirahã Indians of Brazil, have never lost their ability to ensure everyone's freedom and allow everyone to satisfy their needs. Daniel L. Everett provides a fascinating firsthand account of their way of life in *Don't Sleep There Are Snakes: Life and Language in the Amazonian Jungle*. Unfortunately, the Pirahã face a troubled future, because they are surrounded by a rapidly expanding Brazilian population that is destroying the Amazon ecosystem upon which the Pirahã depend.

The Bishnoi of northwestern India is another group that has learned to live well. They conduct their lives by 29 rules established some 500 years ago. These rules include "be compassionate to all living beings," "fell no live trees," and "eat no flesh." Thanks to their ethical code, the Bishnoi have managed to create a sustainable way of life not only for themselves, but for all the wildlife living near their villages. The Bishnoi stand in stark contrast to the other peoples of Rajasthan, whose heedless practices have brought about extensive desertification and local extinctions of wildlife.

Notes

1. Lao Tzu, *The Tao Te Ching of Lao Tzu*, trans. Brian Browne Walker (New York: St. Martin's Griffin, 1995), chapter 80.

2. YouTube video of Edward Abbey giving a talk at the University of Utah. See http://www.youtube.com/watch?v=FZMqjkP8HDM&feature=relmfudward.

3. Aldous Huxley, *Brave New World Revisited* (New York: HarperCollins Publishers, 1958), 118.

4. Leopold Kohr, *The Breakdown of Nations* (Devon: Green Books Ltd., 1957, reprinted 2012), chapter 7, 129.

5. Martin J. Blaser, *Missing Microbes: How the Overuse of Antibiotics is Fueling our Modern Plagues* (New York: Picador, 2014), 47. When the first cities were established in China and the Middle East, new epidemics appeared: smallpox (from cows), measles (from dogs or cattle), influenza (from pigs), and bubonic plague (from fleas borne by rats).

Adjusting to a Larger Proportion of Old People

In a population that is stable or shrinking, the proportion of old people will always be larger than in a growing society. On the plus side, this results in cost savings from less crime and from not having to raise and educate so many children. On the negative side, having more old people means higher expenditures for health care and retirement pensions. Japan and many European nations are already facing the need to increase taxes, reduce benefits, or postpone retirement to compensate for longer lifetimes and fewer younger workers (as well as to pay for ever more expensive medical technology). On the other hand, when families are small, as in Europe and Japan, tax increases can be borne much more easily. And in countries that lack government-provided health care and pensions, parents with few or no children can save more for their own retirement and health care. Another advantage of having more elderly people is that they generally have a smaller ecological footprint than young people, because they tend to live more simply, consuming fewer resources and producing less pollution. In order to generate more tax revenue to support the elderly, some have proposed importing young workers. However, this would not solve the problem, because these workers would also grow old. Worse still, it would increase population size—

especially because immigrants tend to have higher birth rates than natives.

The best way to reduce the cost of supporting more old people would be to improve the population's diet. This would slash costs for medical care and enable the elderly to remain employed longer, either full time or part time. Longer employment would make up for the economic cost of keeping people alive for more years. The number of hours worked per week could be gradually reduced as people age. For those who choose to retire early, Social Security payments would be reduced correspondingly, just as they are today. There is no danger that this would foster job competition between the old and the young, because a shrinking population would create high demand for everyone's labor. Keeping the elderly employed longer would also preserve knowledge within organizations ("institutional memory") and keep the elderly financially solvent. Obviously, not all jobs would be suitable for the elderly: those involving hard physical labor would be inappropriate. But there is one job for which many of the elderly would be ideally suited: teaching. With a lifetime of experience to draw upon, they have much to offer the young as teachers, tutors, mentors, and counselors. Another natural job for the elderly would be to take care of their own grandchildren while their adult children work. This is standard practice in traditional societies, but has become increasingly uncommon in the United States, as parents and children move far away from one another. This geographic separation forces working parents to rely on expensive commercial day care.

The importance of keeping elderly men and women productive was underscored by John Stuart Mill:

There is nothing after disease, indigence, and guilt so fatal to the pleasurable enjoyment of life as the want of a worthy outlet for the active faculties… There are abundant examples of men who, after a life engrossed by business, retire with a competency to the enjoyment, as they hope, of rest, but to whom, as they are unable to acquire new interests and excitements that can replace the old, the change to a life of inactivity brings ennui, melancholy, and premature death. Yet no one thinks of the parallel case of so many worthy and devoted women, who, having paid what they are told is their debt to society—having brought up a family blamelessly to manhood and womanhood—having kept a house as long as they had a house needing to be kept—are deserted by the sole occupation for which they have fitted themselves; and remain with undiminished activity but with no employment for it…[1]

An outstanding example of someone who continued to work far beyond the normal retirement age is Frances Kelsey, a regulatory pharmacologist with the US Food and Drug Administration. In 1962, she became famous for refusing to allow US sales of the antinausea drug thalidomide, despite great pressure from the drug's manufacturer. Her steadfastness prevented thousands of American babies from being born without arms and legs. Frances Kelsey continued to do outstanding work at the FDA until her retirement in 2005 at the age of 90. (She died in 2015.)

Notes

1. John Stuart Mill, *The Subjection of Women* (1869).

National Defense with a Smaller Population

The varieties of man seem to act on each other in the same way as different species of animals—the stronger always extirpating the weaker.

—Charles Darwin[1]

National defense is a critical concern for any country with a shrinking population and shrinking economy. No country can afford to let its population shrink to the point that it can no longer defend itself. Unsustainability can arise from having too few people as well as too many. John Stuart Mill recognized this danger:

> I cannot overlook the many disadvantages to an independent nation from being brought prematurely to a stationary state, while the neighboring countries continue advancing.[2]

Humans seem to be genetically primed for violence. Like chimpanzees, many people find violence exciting:

> Almost all of us long for peace and freedom; but few of us have much enthusiasm for the thoughts, feel-

ings and actions that make for peace and freedom. Conversely, almost nobody wants war or tyranny; but a great many people find an intense pleasure in the thoughts, feelings and actions that make for war and tyranny.

—Aldous Huxley[3]

Edward Abbey offered personal observations on these thoughts and feelings that make for war and tyranny:

Sometimes I frighten myself, the way I slip so easily into irrational anger, blind misanthropy, provincial prejudice. Sometimes I sound like a Nazi or a Jesuit, or a Communist (standard brand), or a Republican congressman from Indiana. It's so pleasant, though, to satisfy the lyric and romantic impulse to make wild and savage and extreme accusations— so much easier than the other course, the only one I respect and trust. I mean the cool rational liberal approach—the balancing of argument and fact, the reasoned and qualified and documented statement, the honest suspension of judgement in case of insufficient, contrary, contradictory or doubtful evidence; the empirical investigation guided by strategic hypothesis and theory.[4]

The proclivity of human males for anger and violence stems from a genetic predisposition inherited from the ancestor we share with chimpanzees:

The evidence suggests that a series of genetic mutations occurred in our ape-like, forest-living forebears, which predisposed adult males to band together with their brothers and cousins to raid and kill their neighbors, and that those who manifested such a trait acquired more territory. More territory meant more resources, more resources meant more females, more females meant the opportunity for more sex, and more sex meant more offspring carrying the male's genes, aggressive tendencies and all, to the next generation. Those males who coordinated their violence in teams became the winners in the ruthless war of nature.[5]

Scientists have so far identified 10 genes that contribute to violent behavior. One of the most studied is the low-activity variant of the MAOA gene.[6] This variant lowers the level of monoamine oxidase A in the brain. This, in turn, lowers the level of the neurotransmitter serotonin. A low concentration of serotonin contributes to a wide range of antisocial and violent behaviors. In New Zealand, the low-activity variant of the MAOA gene is twice as common among Maoris as among those of European descent.[7] Natural selection seems to have favored this gene in the Maoris, presumably because it made them more aggressive, enabling them to conquer and destroy the New Zealanders who preceded them. But once the Europeans arrived, this variant suddenly became maladaptive, because it contributed to a high Maori crime rate. Today, although Maoris comprise only 15 percent of New Zealand's population, they represent 51 percent of its prison population. However, the fact that this gene variant is only twice as frequent in the Maoris, but the Maori crime rate is more than three times greater, shows that there are additional factors that contribute to the high Maori crime rate.

Rapidly growing societies are more prone to violence than demographically stable societies. This is partly because growing populations have shrinking resources, which leads to greater competition. But another factor is the higher proportion of young people in growing societies, which means a higher mean testosterone level. Testosterone plays a key role in both aggression and population growth itself.

There is not much we can do about our genes or the amount of testosterone circulating in our society (although the two-credit system would help in both respects), but as individuals we can certainly try to be more friendly and diplomatic:

> The handshake, the bow, the polite form of speech, and "the soft answer that turneth away wrath" constitute a vital part of the techniques of conflict management, yet to my knowledge they have never been recognized as such or given the importance in human history which they deserve.
>
> —Kenneth Boulding[8]

We could also reduce violence by thinking more carefully. This would allow us to discover and correct our mistakes—as scientists do. Careful reasoning would also make us more aware of our social and cultural conditioning, which would loosen our identification with groups, such as our nation, political party, social class, religion, race, family, even our own species. We might even learn to see ourselves in others.[9]

The role of social conditioning in creating a culture of aggression has been shown experimentally with two species of macaque: the rhesus monkey and the stump-tailed monkey:

> Juveniles of both species were placed together, day and night, for five months. These macaques having strikingly different temperaments: rhesus are a quarrelsome, non-conciliatory bunch, whereas stumptails are laid-back and pacific. I sometimes jokingly call them the New Yorkers and Californians of the macaque world. After a long period of exposure, the rhesus monkeys developed peacemaking skills on a par with those of their more tolerant counterparts. Even after separation from the stumptails, the rhesus showed nearly four times more friendly reunions following fights than is typical of their species. These new and improved rhesus monkeys confirmed the power of conformism.[10]

Most of the world's people expose themselves to images and sounds of violence every day as they watch TV shows, movies, video games, and the evening news. This naturally makes the world seem more dangerous than it really is. If people would turn off their electronic devices and spend more time in nature, they would quickly discover that the world is, for the most part, a very peaceful and pleasant place.

One thing that would help create a more peaceful world would be to recognize the right of all peoples to self-determination. If Kurds, Palestinian Arabs, Basques, Catalonians, Tuaregs, Darfuris, Tigrayans, Chechens, Circassians, Abkhazians, Ukrainians, Tibetans, Uighurs, Taiwanese, Acehnese, Kashmiris, Sikhs, Somalilanders, Puerto Ricans, or any other people wish to govern their own affairs, more power to them. The fair distribution of property and resources following a national divorce or annulment could be handled by an international judiciary of economists and ethicists. Given the opportunity to choose in a referendum, most ethnic groups will favor a substantial measure of autonomy rather than full independence—just as the people of Quebec and Scotland did. Even when people opt for full independence, as in South Sudan or Kosovo, it is still in their best interest to maintain close economic and political ties with their neighbors.

If the principle of self-determination had been respected in the United States 160 years ago, there would have been no Civil War. The North would have respected the right of the Southern states to secede, and the South would have respected the right of slaves to secede from their masters. It was the failure of white Southerners to extend the principle of self-determination to others that precipitated the Civil War. Even today, there are many in the United States who think that freedom means the freedom to exploit and abuse others.

A viable alternative to secession would be to follow the Swiss model of dividing the nation into small cantons, so that no one canton can dominate the others. By means of its canton system, Switzerland has managed to maintain unity and peace for many centuries. If it had not developed the canton system, Switzerland would inevitably have split into three states: German, French, and Italian. And each of these would eventually have merged with its linguistic motherland.[11]

Any serious discussion of national security must address the greatest source of insecurity: nuclear weapons. These horrific weapons were developed by the United States in order to defeat the empires of Germany and Japan. By the time they were ready for use, Germany had already surrendered, and the Japanese were attempting to work out terms of surrender. At that point, the United States should have ceased its aggression against Japan. That would have given the Japanese time to lick their wounds and ponder their folly. Instead, the US dropped two atomic bombs on Japanese civilians, killing approximately 200,000. By comparison with the attacks on the Twin Towers in 2001, the atomic bombings were more than 65 times deadlier. They made the Japanese into victims of war crimes, and that is one reason the Japanese have been slow to acknowledge their own war crimes.

> Skilled command attains, then stops, rejecting force.
> —*Tao Te Ching*[12]

After the war, a nervous Soviet Union and China hastened to develop their own nuclear weapons. This resulted in the Cold War: a 40-year standoff of fear. During this period, the US and the Soviet Union each acquired a few thousand hydrogen bombs.[13] The Cold War nearly became a nuclear war during the 1962 Cuban Missile Crisis. But eventually the Cold War ended when the Soviet Union broke apart and China embraced state-sponsored capitalism. Almost overnight, our sworn enemies became our trading partners, and many of their citizens began immigrating in large numbers to the United States. Unfortunately, there are now once again rising tensions with both Russia and China.

The appeal of nuclear weapons is the promise of safety from aggression. Armed with nuclear weapons, nations know that their neighbors will be afraid to attack them lest they suffer nuclear retaliation. Lost in this calculus is any regard for morality. Nuclear warfare, after all, means deliberately killing vast numbers of innocent people and animals. To use these weapons to retaliate would be as immoral as using them in a first strike—no better than a cornered gunman killing hostages out of spite when his enemy closes in. There is no ethical system in the world that could condone such behavior. Mutually assured destruction would only be ethical if it were applied to leaders rather than populations. If, for example, we required each of the world's leaders to always wear an explosive vest that any other leader could detonate at any time, and if these explosive vests were programmed so that the one who pushed the button to kill would simultaneously blow himself up, then we would have a system of mutually assured destruction that would preserve the peace and spare both civilians and soldiers—and it would cost very little. Unfortunately, persuading our leaders to adopt it would be hard. They would much rather keep the explosive vests strapped on us.

The United States unilaterally created nuclear weapons and should now take the lead in eliminating them. Instead of basing our national defense on the threat to run amok and kill everyone, we should work with other countries to develop an international arms control system—one with an effective police and judicial system that will prevent powerful countries from abusing weaker ones. Those who fear the idea of nuclear disarmament ought to consider that most of us go about our daily lives unarmed, despite the fact that every person we meet could easily kill us. We rely on moral education and a criminal justice system to create deterrence. In England, disarmament extends even to police officers, most of whom do not carry firearms.

Something else we could do to promote peace would be to ban military conscription. Without large armies of conscripts, nations would find it difficult or impossible to occupy other nations. The 13th Amendment of the US Constitution, adopted in 1865, outlawed involuntary servitude except as a punishment for crime, yet the US government impressed young men into the army during WWI, WWII, and during the Cold War until 1973. Even today, the US government contends it has this right, demanding that all 18-year-old males register for involuntary military servitude.

Governments that reject conscription must rely instead upon voluntary recruitment. Their recruitment campaigns depend upon the existence of a class of relatively poor people who see military service as their best hope for improving their economic status. As Malthus observed:

> A recruiting serjeant always prays for a bad harvest and a want of employment, or, in other words, a redundant population.[14]

Another measure that would help to promote peace would be for women to assume a much larger role in governing.

> There has never in the history of humankind been *one example* of women banding together to wage war on another society to gain territory, resources, or power. Think about it. It is always men. There are about nine male murderers for every one female

murderer. When it comes to same-sex homicides, data from twenty studies show that 97 percent of the perpetrators are male.

—Adrian Raine[15]

Women, however, are not saints: female rulers have waged wars and committed atrocities. Nevertheless, female aggression is usually expressed in nonlethal ways, such as gossip and social shunning. Our cousins, the bonobos, have female-led societies. They settle their disputes by giving one another pleasure rather than resorting to violence.[16] Women have 13 to 17 percent more gray matter in their orbitofrontal cortex than men. Some researchers think this anatomical difference may be a key factor in the lower levels of violence in women.[17] Another undoubted factor is the lower level of testosterone in women.

It is often said that people (and their nations) should negotiate from strength. I disagree. Negotiation from strength is not negotiation but imposition. I think successful negotiation comes from mutual vulnerability. I learned this from an encounter I had with a coyote caught in a leghold trap. I was with two companions when we came upon her one December morning at the edge of a farm field. She was terrified, thrashing frantically and putting on a threat display with jaws wide open and barking. After deliberating how we could free her, and coming up with no good solutions, I asked my two companions to step away so the coyote would be less intimidated. I then got down on my belly and attempted to talk reassuringly to her (not looking directly at her). When I began crawling toward her, she panicked, so I stopped. I told her I hated it as much as she did, and that I was on her side. At that point, she somehow understood and stopped resisting. She let me crawl up to the trap and remained still while I figured out the mechanism and released her. Her leg was cut, but not too badly, and she instantly bolted away. The outcome was successful for both of us because we were both vulnerable and both desperate (she to be free and I to free her), so we took a chance on each other.

> One wins by willingly taking the lower position. The other wins by willingly acknowledging its lower position.
>
> —*Tao Te Ching*[18]

A critical, but often overlooked, element of national defense is the need for governments to retain the loyalty of their citizens. The swift collapse of the Afghan government in 2021 and the long-ago collapse of the Roman Empire came about because their citizens preferred to take their chances with the invaders rather than continue to put up with their corrupt, incompetent leaders. To prevent this from happening in our society, we need leaders who are competent and honest (rather than fools and kleptocrats) and an economy that distributes resources equitably.[19]

As more countries adopt population-stabilization plans, the social and environmental stresses that contribute to international hostilities will greatly diminish. When children are wanted, loved, and assured the necessities of life, they are far less likely to become hateful and violent. As the world's population falls, and as women attain more power, and people embrace worthwhile values (including genuine democracy), reliance on armies will fade away.[20]

In ancient days, the good soldier was not violent.
The good fighter indulged no rage. Skilled conquerors engaged no enemy.

—*Tao Te Ching*[21]

Notes

1. Charles Darwin, *The Voyage of the Beagle*, chapter 19.

2. John Stuart Mill, *Principles of Political Economy* (New York: Prometheus Books, 2004), 806. Originally published in 1848.

3. Aldous Huxley, *Brave New World Revisited* (New York: HarperCollins Publishers, 1958), 50.

4. 4. Edward Abbey, *Confessions of a Barbarian* (Boulder, CO: Johnson Books, 1994), 90. Journal entry of June 9, 1952.

5. Malcolm Potts and Thomas Hayden, *Sex and War: How Biology Explains Warfare and Terrorism and Offers a Path to a Safer World* (Dallas: BenBella Books, 2008), 12.

6. 6. Adrian Raine, *The Anatomy of Violence: The Biological Roots of Crime* (New York: Vintage Books, 2013), 50–54.

7. Ibid., 54–55.

8. Kenneth Boulding, *The Meaning of the 20th Century: The Great Transition* (New York: Harper Colophon, 1964), 95.

9. One of the uglier aspects of group membership is cruel initiation rites like those practiced by the military, fraternities, religions (circumcision, animal sacrifices), and some street gangs (which require a murder for membership).

10. 10. Frans de Waal, *Are We Smart Enough to Know How Smart Animals Are?* (New York: W. W. Norton & Company, 2016), 256. Frans de Waal is a primatologist.

11. Leopold Kohr, *The Breakdown of Nations* (Devon: Green Books Ltd., 1957, reprinted 2012), chapter 3, p. 76. Today, Switzerland faces a new challenge in the form of Muslim immigration. The Swiss Muslim population has risen from 1 percent in 1980 to more than 5 percent today. However, most of Switzerland's Muslims are ethnic Slavs, not people from the Middle East. Their ancestors converted from Christianity to Islam when Bosnia and Kosovo were under Turkish rule.

12. Lao Tzu, *Tao Te Ching: A New Translation*, trans. Sam Hamill (Boston: Shambhala Publications Inc., 2005), chapter 30.

13. The United States currently has about 7,200 nuclear warheads. Russia has 7,500; France 300; China 260; UK 215; Pakistan 120 to 130; India 110 to 120; Israel 80; North Korea fewer than 10. The manufacture of these weapons has resulted in long-term plutonium pollution and high rates of cancer in the vicinity of weapons-manufacturing facilities. Plutonium-239 has a half-life of 24,100 years.

14. Donald Winch, ed., *An Essay on the Principle of Population* (Cambridge University Press, 1992), 222.

15. Adrian Raine, *The Anatomy of Violence: The Biological Roots of Crime* (New York: Vintage Books, 2013), 33. Raine's words echo those of Malcolm Potts and Thomas Hayden: "This is perhaps the most profound insight to come from taking an evolutionary perspective on war: empowering women reduces the risk of violent conflict. Far from being a politically correct notion of feminist philosophy, women's role in reducing the risk of war is borne out by rigorous study and historical experience" (*Sex and War*, p. 14).

16. Frans de Waal and Frans Lanting, *Bonobo: The Forgotten Ape* (Berkeley: University of California Press, 1997).

17. Adrian Raine, *The Anatomy of Violence: The Biological Roots of Crime* (New York: Vintage Books, 2013), 151.

18. Lao Tzu. *The Tao Te Ching of Lao-Tzu*, trans. Brian Browne Walker (New York: St. Martin's Press, 1995), chapter 61.

19. Egalitarian nations fare much better in warfare than hierarchical nations: "After examining two dozen recent conflicts researchers found that the nation with the more equitable distribution of wealth defeated the nation with the less equitable distribution in three-quarters of the wars. When they looked at eighty wars going back to Napoleon, they observed that the nation with the least social stratification won in eight out of ten conflicts" (*Sex and War*, p. 190).

20. One part of the world where there is a critical need to relieve the social and environmental stresses brought on by overpopulation is Palestine/Israel. Currently, Arabs and Jews are engaged in a race to outbreed each other. This conflict could be ended easily and permanently if both sides would agree to reduce their respective populations to a small, reasonable number—say 50,000 Arabs and 50,000 Jews for the whole of Palestine (more than enough people for such a tiny, arid, exhausted land). In order to reach that final size, they should adopt a two-credit system—one with strict and mutually verifiable enforcement provisions. This program could be supervised by a team of demographers from the United Nations. Each side's rate of population reduction would be adjusted so that both sides would reach the final goal at the same time. Once their birth rates start to fall, the two sides would be allowed to receive immigrants from their respective diasporas provided they compensated for this by further lowering their birth rates. The equitable distribution of land, water, and other resources could be handled by an international body of jurists, ecologists, economists, and ethicists. The borders shared with other states (Egypt, Jordan, Syria, Lebanon) could be patrolled by a small force of UN peacekeepers. The first task of the peacekeepers should be to disarm both sides, starting with the Israeli nuclear weapons. To help break down religious and linguistic barriers, all Jewish and Arab children should attend secular summer camps together. As the Palestinian Arab and Jewish populations shrink, the supply of surplus land will greatly increase—especially

if limits are placed on how much land individuals and corporations can own. This surplus land should be permanently set aside as parkland. Both sides can then devote themselves to restoring their ravaged environment. They can eliminate all the herds of sheep and goats, restore the drained wetlands (including former Lake Hula), and plant millions of trees to reestablish long-vanished woodlands. Habitat restoration will moderate the local climate and eventually allow gazelles, deer, ibex, oryx, wild boar, cheetahs, leopards, lions, Syrian bears, and wolves to once again share the land with people. A new Garden of Eden will replace the threat of Armageddon.

21. Lao Tzu, *Tao Te Ching: A New Translation*, trans. Sam Hamill (Boston and London: Shambhala Publications Inc., 2007), chapter 68.

Reversing Climate Change

To restore our climate to its preindustrial state, we must remove 33 gigatons of excess carbon from the atmosphere. Most of this excess carbon will be reincorporated into the soil or stored in woody plants. What cannot be returned to the soil or stored in plants will gradually be absorbed by the oceans (which have always stored the bulk of the world's carbon).[1]

There is a direct relationship between population size and the amount of CO_2, methane, and nitrous oxide emitted. Cutting our population size in half would automatically cut our harmful emissions in half. And with half as many people, only half as much land would need to be cultivated. The surplus land would quickly revert to its original ecosystem of forest, grassland, or wetland. As vegetation reclaimed this land, atmospheric CO_2 would be captured by photosynthesis and stored in woody plants or incorporated into the soil (as topsoil, peat, and muck). Reducing population size would therefore attack the problem of climate change on two fronts: *prevention* (cutting emissions) and *treatment* (capturing and storing $CO2$). In the US (which is the largest emitter of climate-changing gases after China), the quickest way to reduce population size would be to halt mass immigration, which currently accounts for 88 percent of US population growth.

Another powerful way we could slash emissions of CO_2 while simultaneously capturing and storing a large amount of CO_2 would be to switch from a diet heavily based on animal products to one based on plants. The adoption of a plant-based diet would end all the CO_2 emissions that come from feeding, housing, transporting, refrigerating, and cooking hundreds of millions of animals annually. It would also eliminate the large quantities of methane emitted from their digestive tracts. (Methane is 21 times more potent as a global warming gas than CO_2.) Moreover, adoption of a plant-based diet would capture existing CO_2 by removing from production a vast amount of farmland that is currently used to grow food for farm animals. This land would quickly revert to woodland, prairie, or wetland. According to the 2012 US National Resources Inventory (published in 2015), 19 percent of US land cover consists of cropland, and 6 percent consists of pastureland (as distinct from Western rangeland). If the American people adopted a plant-based diet, much of America's pastureland and at least half of its cropland would revert to forest and wetland. This represents at least 16 percent of the total land cover of the United States. This percentage could be further boosted by abandoning the use of corn to produce alcohol. The area of the US currently devoted to corn production is 34,254,507 hectares (nearly the size of California). Of this, 40 percent is used for ethanol production, and 36 percent is used for animal feed. The conversion of corn to ethanol is an inefficient way to obtain fuel (see the energy-returned-on-energy-invested table later in this chapter), and it is only economically viable because farmers receive government subsidies to grow corn. Eliminating these subsidies would lead to the abandonment of much of the current corn cropland, allowing it to revert to forest or another natural ecosystem. Moreover, if people switched to a plant-based diet, demand for ethanol and fossil fuel would fall, because the production of plant protein requires 8 times less energy than production of animal protein.

Among the other benefits of reducing the amount of land under cultivation would be a reduction in the rate of soil erosion. In fact, the rate of soil erosion would decrease more than proportionally, because farmers would cease cultivating hillsides in favor of the richer flatlands. The quantity of fertilizers, herbicides, and pesticides applied to the land would also plummet. And less use of fertilizer would result in fewer and smaller algal blooms in our lakes and oceans. Not only our environment but we ourselves would be much healthier if we adopted a plant-based diet. The incidence of heart attacks, strokes, diabetes, and cancers would fall rapidly. This would save the nation vast sums of money currently spent on health care. It would also move us up a few notches on the moral scale if we weren't exploiting and killing our fellow beings. Yet another advantage to adopting a plant-based diet is that it would lower food prices. This is true for two reasons: (1) it is far cheaper to eat plant foods directly than to first feed them to animals and then eat the animals and their milk products; and (2) the land that farmers would abandon would be their least productive land, allowing food to be grown only on the most productive land, thereby lowering the cost of production.[2]

The reversion of farm fields to forest has already taken place in some parts of the world. In New England, for example, forests now cover much of the land that was farmed in the 18th and 19th centuries. Reforestation also occurred long ago in Central America following the collapse of the Mayan city-states and in Cambodia following the collapse of the Khmer Empire.

Another effective way to capture and store carbon is to restore organic matter to our agricultural soil. This will require putting an end to plowing and synthetic fertilizers. Instead, farmers will need to use cover crops and regularly fallow their fields, just as their predecessors did. This process of soil regeneration will quickly pull carbon out of the atmosphere and return it to the soil. It will also make the soil far more fertile and less toxic. And it will protect our rivers, lakes, and oceans from the massive amounts of nitrate and phosphate fertilizer that currently washes off farm fields.

Reducing population size and reducing our intake of animal flesh would not only restore our carbon-storing soil, forests, and grasslands, but would halt the collapse of the world's fish populations. One of the little-known but critical ecological functions performed by fish is to nourish carbon-storing coral reefs and marine algae with their excrement. Another important function of fish is to mix warm surface water with cooler deep water, thereby cooling the near-surface waters where most marine life resides. Fish accomplish this through the movements of their schools up and down in the water column. Another reason to reduce fish consumption is the criminal nature of much of the world's commercial fishing. It is estimated that one-third of the fish sold in the US and other countries is illegally caught. And on the high seas there is no law at all. Moreover, many of the crews on Southeast Asian fishing boats are enslaved. The various organizations that claim to certify brands of seafood as "sustainable" never do inspections. They sell their bogus certificates like medieval priests selling indulgences to sinners.

Attention is increasingly being drawn to the vast quantities of plastics polluting our oceans and entering the flesh of marine animals (and all who eat them). What is rarely mentioned is that a significant source of this plastic is abandoned fishing gear. Abandoned nets not only cause plastic pollution, but they continue to kill marine life as they drift about in ocean currents. Most environmental organizations don't want to talk about this, because they receive large donations from the fishing industry. A notable exception is the Sea Shepherd Conservation Society, which has fought long, hard, bravely, and effectively against illegal fishing and whaling.

Eating shrimp, by the way, is no better than eating fish: the world's shrimp industry has destroyed a large portion of the world's critical mangrove forests, which store carbon, protect against storm surges, and serve as nurseries for marine life. Moreover, commercially farmed shrimp are heavily polluted. In fact, *all* marine life is polluted with mercury, PCBs, dioxins, plastics, and other effluents of our soulless way of life. A cause for hope is that lab-grown seafood is now being produced experimentally in Singapore. Once production costs for these lab meats become lower than the costs of harvesting live animals, the lab meats will take over the market. Governments could expedite this transition by subsidizing lab meats.

In addition to reducing population size, and halting our consumption of terrestrial and marine flesh, we can reduce climate change by reducing our personal consumption of goods. If we cut our consumption of material goods in half, we would eliminate half of the climate-changing gases emitted during their production, transportation, and subsequent use (as well as save resources for future generations).

Nothing could be more salutary at this stage than a little healthy contempt for a plethora of material blessings.

—Aldo Leopold[3]

We could further reduce CO_2 emissions by purchasing products that are durable, made of recyclable materials, and made with less material and energy.

> If the average lifetime of each product flowing through the human economy could be doubled, if twice as many materials could be recycled, if half as much material needed to be mobilized to make each product in the first place, that would reduce the throughput of materials by a factor of eight.[4]
>
> —Donella Meadows, et. al.

Two nations are responsible for nearly 50 percent of the world's climate-changing gases: China and the United States. China releases more CO_2 than the US, but a substantial share of this is due to producing goods for American consumption. Efforts to reverse climate change obviously need to be concentrated on those two nations.

So far, I haven't even mentioned the two most widely entertained responses to climate change: cleaner production of electricity and cleaner transportation. There was really no need to mention them, because reducing population size would automatically slash emissions of climate-changing gases by power plants and vehicles. This process could be further accelerated by adopting more fuel-efficient vehicles, establishing efficient mass transportation systems, electing politicians who will provide subsidies for installing solar panels on homes and businesses and require power companies to install more wind turbines and solar arrays. As a temporary measure, all coal-burning plants should be converted to natural gas. (The burning of natural gas produces no toxic ash and lowers CO_2 emissions by 50 percent.) Demand for energy could also be reduced by time-of-day pricing for electricity, which encourages consumers to use electricity during hours when demand is low. We could also pass laws requiring motion detectors for outdoor lighting systems such as those in parking lots, so that lights would come on only when moving vehicles or people were sensed. This would not only save energy and money but reduce light pollution for those who enjoy looking at the night sky.

International shipping emits as much CO_2 as all of America's coal-burning power plants.[5] The world's nations should impose a sliding tax on ships that dock at their ports depending on how polluting they are. Nations should also convert ships to burn natural gas or, better yet, use hydrogen fuel cells. Modern sail technology could also significantly augment ship efficiency.

The following table shows the percentage of energy returned on energy invested (EROEI) for various energy sources.

Power Source	EROEI
Hydroelectric	100
Coal	80
Thin-film solar	Up to 60 in US Southwest
Nuclear with centrifuge enrichment	50-75
Monocrystalline silicon solar panels	19-39
Wind turbines	18-20
Oil and gas	12
Nuclear with diffusion enrichment	10
Ethanol from sugarcane	5
Ethanol from corn	<2 (and perhaps zero or negative)

The EROEI table shows that hydropower is by far the most efficient way to produce power. It is also clean, except for the CO_2 emitted during the manufacture of the dam's cement and turbines and that emitted by construction equipment. But dams inundate beautiful valleys, destroy migratory fish populations, dry up estuaries, and displace wildlife and humans. Dams currently supply about 6 percent of America's electricity, but in the Pacific Northwest they supply more than 50 percent.[6] All hydroelectric reservoirs eventually fill with silt, eliminating the dam's capacity to store water and generate electricity.

Nuclear power, although it does not contribute climate-changing gases, has three critical unsolved problems: (1) susceptibility to radiation leaks (Chernobyl, Fukushima), (2) absence of any secure way to store highly radioactive spent fuel for thousands of years, and (3) the ease with which nuclear fuel can be processed to produce nuclear weapons. Another drawback of nuclear power is its expense.

It is never economical if we take into account the cost of securely storing nuclear waste for thousands of years, the cost of deactivating obsolete power plants, the cost of trying to clean up radioactive contamination (Chernobyl, Fukushima), and the long-term loss of land from contamination. In the US, nuclear power plants supply about 20 percent of the nation's electricity,[7] while in France they supply 75 percent. Worldwide the figure is about 15 percent. Most nuclear power plants are old and will soon have to be retired at great expense—an expense that will continue for generations to come.

However, the recent development of small nuclear reactors may yet make nuclear power practical. These small reactors are relatively inexpensive because they are factory-produced. In addition, they don't need to be refueled as frequently as large power plants and their fuel cannot be made into weapons. It remains to be seen whether they can compete economically with wind and solar.

The use of natural gas has its own set of problems. Although power plants that burn gas emit only half as much CO_2 as those that burn coal, they are still huge emitters of CO_2. Natural gas production is also problematic. The process of extracting natural gas by hydraulic fracturing ("fracking") releases large amounts of global warming gases (including ethane) and pollutes water with toxic chemicals. Fracking also triggers earthquakes, which has become a major problem in Oklahoma. Another drawback of natural gas is leakage from underground storage reservoirs. In 2016, a leaking gas storage facility at Porter Ranch outside Los Angeles released each day an amount of methane equivalent in global warming effect to driving 4.5 million cars for a day.[8]

Unlike wind turbines, which must be concentrated in areas that have strong and reliable winds, solar power can, and should, be decentralized. Solar panels should be installed on rooftops and over parking lots. Existing battery technology already allows solar power to be stored for nighttime use. By decentralizing the production of electricity, we could eliminate power blackouts (whether accidental or malicious) and eliminate the excessive political power of the utility companies and the coal, gas, petroleum, and nuclear industries. The current Russian missile attacks on the Ukrainian power grid demonstrate why power production should never be centralized. Yet our unthinking politicians call for expanding and improving the power grid in order to accommodate distant wind farms and solar arrays.

In the future, we will have less energy available and a much smaller economy. We will eventually have to learn to live on the daily revenue provided by the sun (which includes wind) rather than drawing down solar capital that was stored in the earth 300 million years ago. If we're smart, we will make this transition voluntarily and not wait for nature to force it upon us.

It is remarkable that current public discussions about climate change ignore its link to population size.[9] There is deep reluctance to acknowledge that the multiplier of fossil fuel consumption, forest destruction, aquifer depletion, and soil erosion is human numbers. Even if we could achieve net-zero carbon emissions today, that would do little to halt the destruction of natural areas, the extinction of plants and animals, the erosion of topsoil, and the overcrowding that generates hunger, violence, and epidemics. But reversing population growth would eliminate *all* those harms *and* eliminate climate change.

America's profligate use of energy resources began with the first settlers. Peter Kalm, a Swedish botanist who traveled throughout eastern North America from 1748 to 1750, observed in amazement that many Americans kept several fires burning in their homes (something no sensible Swede would do).[10] Thoreau, too, remarked on the way his neighbors wastefully burned wood, thereby causing deforestation around his hometown of Concord, Massachusetts:

> A sermon is needed on economy of fuel. What right has my neighbor to burn ten cords of wood, when I burn only one? Thus robbing our half-naked town of this precious covering. Is he so much colder than I? It is expensive to maintain him in our midst.[11]

Notes

1. Atmospheric CO_2 in the Northern Hemisphere has an annual cycle: The level is highest in May and lowest in September. It falls during the growing season as CO_2 (combining with H_2O) is converted by photosynthesis into glucose (simple sugar) and then cellulose (wood). The Southern Hemisphere, being mostly tropical, lacks a distinct growing season. In September 2016, the CO_2 concentration exceeded 400 ppm for the first time since before the Ice Age. The storage of CO_2 in forests and wetlands is not keeping pace with mankind's massive release of CO_2.

2. The drop in food prices that follows abandonment of the least fertile lands was discussed by John Stuart Mill in *Principles of Political Economy* (New York: Prometheus Books, 2004, 662–63): "There are two kinds of agricultural improvements. Some consist in a mere saving of labor, and enable a given quantity of food to be produced at less cost, but not on a smaller surface of land than before. Others enable a given extent of land to yield not only the same produce with less labor, but a greater produce; so that if no greater produce is required, a part of the land already under culture may be dispensed with. As the part rejected will be the least productive portion, the market will thenceforth be regulated by a better description of land

than what was previously the worst under cultivation." Though Mill did not mention it, the same abandonment of the least fertile land would follow population reduction.

3. Aldo Leopold, *A Sand County Almanac* (Oxford University Press, 1949), ix.

4. Donella Meadows et al., *Limits of Growth: The 30-Year Update* (White River Junction, VT: Chelsea Green Publishing, 2004), 121.

5. See Matt Apuzzo and Sarah Hurtes, "Tasked to Fight Climate Change, a Secretive U.N. Agency Does the Opposite," New York Times, June 3, 2021: https://www.nytimes.com/2021/06/03/world/europe/climate-change-un-international-maritime-organization.html?searchResultPosition=2

6. See http://www.eia.gov/energy_in_brief/index.cfm.

7. Ibid.

8. Justin Worland, "The On-going California Natural Gas Leak is a Disaster for the Planet," *Time,* January 25, 2016, 16–17.

9. In a *Time* magazine article dated September 23, 2019, renowned primatologist and environmentalist Jane Goodall declared: "In order to slow down climate change, we must solve four seemingly unsolvable problems. We must eliminate poverty. We must change the unsustainable lifestyles of so many of us. We must abolish corruption. And we must think about growing human population. There are 7.7 billion of us today, and by 2050, the UN predicts there will be 9.7 billion."

Jane Goodall knows very well that overpopulation is the primary cause of poverty and that poverty contributes to corruption. Yet, like so many environmentalists, she carefully couches her words about overpopulation to avoid offending donors. It is telling that the first three items on her agenda are calls to action—*eliminate, change, abolish*—but when it comes to population growth, the best she can do is recommend that we *think* about it.

10. Pehr Kalm, *Peter Kalm's Travels in North America: The English Version of 1770* (Toronto: Dover Publications, 1937, 1964). This is a fascinating account of a botanist's travels in eastern North America in the mid-18th century.

11. Henry David Thoreau, *Journal*, April 26, 1857.

Restoring Ecosystems

The "Rewilding" of Abandoned Land

As population size decreases, most residents of the suburbs will migrate to the cities or take up farming. Abandoned suburban homes will be torn down and their materials recycled. The suburban land will then revert to its original ecosystem of forest, prairie, or desert, resulting in the capture of large quantities of atmospheric carbon.

Much of the rural American West will experience depopulation as residents die off or migrate to more densely populated areas in quest of work and companionship. In the vacated regions of the Great Plains and Great Basin, fencing can be torn down, and the original fauna of bison, pronghorn, prairie dogs, black-footed ferrets, wolves, and grizzly bears can return after an absence of 150 years. Thoreau would rejoice:

> A people who would begin by burning the fences
> and let the forest stand![1]

One benefit of a much smaller population in the arid lands of the western US would be a great reduction in the number of human-ignited wildfires. That would allow the resumption of the natural cycle of infrequent fires ignited by lightning.

Reviving Rivers

Most riparian ecosystems in America have been severely degraded not only by pollution but by dams, levees, diversions, and routing through underground pipes.

Dams provide electrical power, flood control, and provide water for agriculture and cities. They also trap sediment to protect downstream dams whose reservoirs would otherwise silt up more quickly. But these benefits come at a heavy cost. As previously mentioned, they destroy migratory fish populations,[2] inundate beautiful valleys (such as the Hetch Hetchy Valley in Yosemite National Park and the Colorado River's Glen Canyon), and cause estuaries to dry up. The Colorado River estuary was once a vast marsh teeming with wildlife but is now a barren wasteland with only a trickle of water passing through. The great marshes at the mouths of the Tigris and Euphrates rivers are likewise shrinking as ever more water is consumed upstream. River deltas can be harmed not only by reductions in water input but by the loss of sediment input. The fertile Nile delta is being consumed by the Mediterranean Sea because the sediment that used to replenish it is now trapped behind the Aswan High Dam. The good news is that as soon as human populations start shrinking, so will demand for electricity and water, allowing dams to be torn down and riparian ecosystems to be restored.

If we fail to take down the dams, nature will eventually do it for us. Massive amounts of sediment have already accumulated in the impoundments and will eventually fill them. All dams are temporary—as every beaver knows. And once the dams are gone, migratory fish will return:

Perchance, after a few thousands of years, if the fishes will be patient, and pass their summers elsewhere meanwhile, nature will have leveled the Billerica dam, and the Lowell factories, and the Grass-ground river run clear again, to be explored by new migratory shoals, even as far as the Hopkinton Pond and Westborough swamp."

—Henry Thoreau[3]

Levees, like dams, wreak havoc on river ecosystems. By preventing flooding, they prevent fertile sediment from being deposited in the natural floodplain—sediment that formerly nourished rich bottomland forests and created protective marshes near the river's mouth. The channelization of the Mississippi River has resulted in the rapid erosion of Louisiana's coastal marshes, making coastal settlements more vulnerable to hurricane damage. The sediment that once replenished these marshes is now discharged far out in the Gulf of Mexico, where it settles to the bottom. There, its high content of nitrogen and phosphorus (from fertilizer runoff) creates an anoxic dead zone the size of New Jersey. In China, eroded sediment deposited between levees along the Yellow River gradually raised the bed of the river 30 feet above the river's floodplain. When the river breached these levees (as it inevitably did from time to time), mass drowning and famine ensued. The source of all the sediment was upstream erosion caused by farming on deforested slopes. Nowadays, the Yellow River often dries up due to overdrawing of water.

A sensible solution to the problems created by levees would be to redirect the money currently spent for levee construction and maintenance toward building flood-proof dikes around major flood-prone towns and cities. (The Dutch are experts at this.) Once these urban areas are protected, the rural floodplain beyond the dikes should be allowed to undergo natural flooding. The federal government should defray the cost of relocating people from the floodplains, since the government built the levees in the first place (albeit with strong local support). These relocations would be a one-time expense, in contrast to having to repeatedly dole out money to flood victims through the Federal Emergency Management Agency (FEMA). Those who choose to remain in vulnerable floodplains would be wise to live in motor homes, so they could quickly relocate to higher ground when rivers rise. Alternatively, they could build truly flood-proof homes by piledriving a tall steel beam into the ground and then building a house on floats around it. The house would float up and down in response to fluctuating water levels, like a toilet float. As for the barge traffic that currently depends on channelization of the Mississippi, it could easily be replaced by railroad transport. Railroads cause much less environmental harm than does the massive infrastructure that sustains barge traffic. Moreover, it is unfair for the government to subsidize barge traffic while railroads have to pull their own weight.

Streams in urban areas are often invisible because they have been diverted into underground pipes. This eliminates the need to build bridges and culverts and formerly concealed the sight and stench of human sewage. Some communities are now "daylighting" their buried streams, re-creating meandering channels for them, and planting attractive landscapes along their banks. These restored riparian corridors are popular with pedestrians and cyclists and can serve as safe travel corridors for wildlife.

Notes

1. Henry David Thoreau, "Walking," (1861).

2. More than 100 salmon stocks in the Pacific Northwest and California are already extinct. About one-half of the remaining stocks are in danger of extinction.

3. Henry David Thoreau, *A Week on the Concord and Merrimack Rivers* (Boston: Houghton Mifflin Company, 1849), "Saturday," 32. Thoreau was one of the first to decry the devastating impact of dams on migratory fish. Upon observing fish trapped by a dam in the Massachusetts town of Billerica, he contemplated sabotage: "Who knows what may avail a crow-bar against that Billerica dam." Sadly, the Billerica dam still stands. Thoreau's dream of a Concord River running free was echoed in Edward Abbey's final speech, delivered in January 1989 at the University of Utah. Abbey concluded his talk by imagining the Colorado River once again flowing freely through Glen Canyon. His talk can be viewed online. It would have gratified Abbey to know that as of 2022, a considerable stretch of Glen Canyon is once again exposed thanks to a decades-long drought which has left Lake Powell at only 25 percent of its full capacity. Moreover, that capacity is 7 percent less than its original capacity due to the deposition of silt over the past half century.

Making Agriculture Sustainable

Agriculture is an excellent source of revenue for a country with a shrinking population. A relatively small number of farmers can feed the nation and still produce a surplus for export to countries that cannot grow enough to meet their own needs. But American agriculture, as it is currently practiced, is unsustainable. Topsoil is rapidly eroding, aquifers are being drawn down, and rivers, lakes, and oceans are being polluted by fertilizer runoff. One-half of Iowa's topsoil is already gone. So is one-third of the soil of eastern Washington's fertile Palouse region. Over the next 100 years, during a period of unprecedented demand for food, one-third of US cropland—that which occupies the steepest slopes—will cease to be productive due to erosion.[1]

Soil, not oil, is our foremost strategic material. Yet for every ton of grain produced, we lose a ton or more of topsoil to erosion. We are losing topsoil 10 to 40 times faster than it is being replenished. This is a hidden cost, one not paid by the current generation of Americans. We are drawing down our soil capital instead of building it up.

> The USDA estimates that about half the fertilizer used each year in the United States simply replaces soil nutrients lost to topsoil erosion. This puts us in the odd position of consuming fossil fuels—geologi-

cally one of the rarest and most useful resources ever discovered—to provide a substitute for dirt—the cheapest and most widely available agricultural input imaginable.[2]

Every farm should be adding back whatever quantity of organic matter and nutrients it loses to oxidation, erosion, and crop removal. This is not impossible: in the Colca Valley of Peru, the indigenous people have farmed their terraces for 1,500 years—and their soil is still fertile. They achieved this almost unique feat by using crop rotation (including nitrogen-fixing legumes), intercropping, fallowing, and—very importantly—seed drilling instead of plowing.[3] Nothing has done more to destroy the world's soil (and thus more to destroy civilizations) than the plow, because it exposes soil to rapid erosion by wind and water. Sustainable agriculture will depend on putting an end to plowing. Thankfully, the move away from the plow is already well underway: Conservation tillage (which leaves at least 30 percent of the soil covered with crop residue) and no-till techniques were used on 60 percent of Canadian farms by 2001, while in the US, conservation tillage and no-till were used on 41 percent and 23 percent of farms respectively.[4] No-till not only greatly reduces erosion but increases soil fertility. It also reduces CO_2 release (and thus global warming). It does this in two ways: first, by cutting the farmer's fuel use in half, and second, by reducing the oxidation of soil carbon. Unfortunately, no-till agriculture currently depends on the use of selective herbicides to control weeds, but that is still better than losing topsoil. Another helpful step to reduce soil erosion would be to more closely mimic natural ecosystems by increasing our use of polycultures of perennial species such as berry bushes, asparagus, rhubarb, and trees that produce fruit, oil, or nuts. But, of course, we would still have to rely heavily on annual species like wheat, rice, maize, and beans.

With respect to fertilizer, farmers will have to stop using chemical fertilizers and instead use green manures (cover crops). This will rapidly build up the topsoil and improve crop yields. As plants absorb CO_2 and transfer it into the soil, microbes will sequester it as humus. This will regenerate a healthy soil microbiome in only a few years. It is estimated that 20 to 40 percent of the carbon in the atmosphere was originally bound in the soil but was released by bad farming practices.

Although our grain exports bring in useful income, they require massive amounts of water and increase topsoil erosion, while harming the importing countries by encouraging population growth (and thus accelerating environmental degradation). We should never sell or donate food to poor countries without requiring that they practice effective birth control in return. Otherwise, we worsen environmental degradation and abet their ruin. If their overpopulation is due to ignorance or lack of contraceptives, then we should help them with education and IUDs, but we must stop feeding the problem. To allow poor people to "call into existence swarms of creatures who are sure to be miserable, and most likely to be depraved"[5] is far more cruel than requiring them to restrain their numbers as a condition for receiving food.

The critically important Ogallala Aquifer, which underlies several High Plains states, supplies water for about 20 percent of US cropland. It is currently being overdrawn for irrigation at more than 10 times the natural recharge rate.[6] It won't be long until this aquifer is exhausted or the cost of pumping its water exceeds the profit from the crops. The same thing is happening in California's Central Valley, where the land is already subsiding due to water loss. Strong measures should be taken now to prevent this tragedy of the unregulated commons. The nonfarming public could help by reducing or eliminating their consumption of animal products, because most of the Ogallala water is used to irrigate corn, soybeans, and alfalfa, which are fed to animals.

We have an ethical responsibility to transfer our land to future generations unimpaired (and preferably improved). Unfortunately, industrial agriculture cares only about profit, not posterity. It obtains profit by eroding the soil, filling our waterways and oceans with silt and pollutants, and polluting our bodies with pesticides, antibiotics, and hormones. It consumes natural capital instead of living off the income provided by the sun's energy. Industrial agriculture must be replaced with small family farms that support healthy farming communities. These family farms will thrive because farmers will finally receive a fair price for their food. The elimination of industrial farming would also protect small farmers in poor countries by eliminating the mass export of American crops at such low prices that farmers in those countries are ruined and forced off their land, turning them into internal or external migrants. By reducing our population size, we will reduce soil erosion, aquifer depletion, fertilizer runoff, and climate change—even if we take no other measures.

Notes

1. David R. Montgomery, *Dirt: The Erosion of Civilizations* (Berkeley: University of California Press, 2007), 163.

2. Ibid., 200.

3. Ibid., 80.

4. Ibid., 211.

5. John Stuart Mill, *Principles of Political Economy* (New York: Prometheus Books, 2004), 346. Originally published in 1848. Here is the full passage: "…but there is a tacit agreement to ignore totally the law of wages, or dismiss it in a parenthesis, with such terms as 'hard-hearted Malthusianism;' as if it were not a thousand times more hard-hearted to tell human beings that they may, than that they may not, call into existence swarms of creatures who are sure to be miserable, and most likely to be depraved; and forgetting that the conduct, which it is reckoned so cruel to disapprove, is a degrading slavery to a brute instinct in one of the persons concerned, and most commonly, in the other, helpless submission to a revolting abuse of power."

6. Manjula V. Guru and James E. Horne, "The Ogallala Aquifer," The Kerr Center for Sustainable Agriculture, published July 2000, 8, http://www.kerrcenter.com.

Creating Efficient Transportation

As our population declines, so will road traffic and the tax base that supports road construction and repair. Many paved roads will be converted to dirt roads or abandoned altogether. With fewer vehicles on the roads, traffic injuries and deaths will sharply decline—for humans and other species.

For land travel between cities, people will increasingly rely on swift, reliable trains. In contrast to America's current passenger trains, these trains will be given right-of-way priority over freight trains. When people arrive at their destination, they will be able to quickly obtain a car or taxi or electric bike to take them to their final destination.

Those who worry that efficient trains will be too costly, fail to consider the high costs of the current transportation system, including these:

- Adverse climate change[1]
- Air pollution
- Highway construction and maintenance
- Financing, fueling, insuring, and repairing vehicles
- Maiming and killing thousands of people and millions of nonhuman animals every year
- Funding a large navy to protect imports of crude oil.

The good news is that in the US the infrastructure for a modern efficient train system already exists—and is publicly owned. I mean the interstate highway system. By expelling cars and trucks from the interstate highways and laying rails on them, the problem is solved. At least two rail lines would be laid in each direction: one express, the other local. This interstate rail system would be the safest in the world because every line would be unidirectional, and there would be no road crossings (hence no danger of collisions and no need for ear-splitting locomotive whistles and crossing gates). And if the population was shrinking at the same time the rail system was introduced, the additional traffic shunted onto the surface roads would not cause excessive congestion. But even if the population was not shrinking, the resulting traffic congestion would give people a strong incentive to switch to the rail system. Another benefit is that the rails could be set farther apart than is the current standard (thanks to broad freeway shoulders). That would allow rail cars to be substantially wider and more comfortable. Moreover, railroad ties would not be needed. That would help keep carbon stored in trees.

Another benefit of a freeways-to-rails system is that people living in the vicinity of freeways would no longer have to endure the incessant din of car and truck traffic. For the first time they would be able to open their windows and talk to their neighbors. And they would no longer have to breathe polluted air.

A major inefficiency of today's passenger trains is that they must stop at each station to receive and discharge passengers. This greatly increases travel time and consumes much energy as the train repeatedly accelerates to cruising speed. A simple solution would be to build an auxiliary track on both sides of the station that would connect to the main track, allowing a pod car containing the new passengers to accelerate up to the back of the passing train, and connect to it electromagnetically. A moving walkway in the pod car would then quickly transfer passengers into the back of the train, as a parallel moving walkway moved the exiting passengers into the pod car, which would then return to the station. This would not be a difficult technical challenge.

Transferring much of our current truck traffic to railroads would cause road maintenance costs to plummet, because trucks account for about 95 percent of road damage. (Road repair represents a massive public subsidy for the trucking industry.) Trucks also crack tens of thousands of car windshields every year from the stones they kick up. Of course, trucks would still be used for local transport, so the system would be a hybrid one.

Trains operating on the former interstate highway system should be powered electrically. This would be quieter, cleaner, and cheaper than diesel locomotives. To prevent trains from colliding with obstacles or derailing due to heat-buckled tracks, bad switches, or sabotage, webcams should be located along every stretch of track so the locomotive engineer (assisted by an automated safety system) can detect any upcoming hazard well in advance.

This train system should remain publicly owned in order to ensure that all communities currently served by the 43,000 miles of interstate highways would be served by high-speed train service. Any excess capacity can be rented out to freight companies.

In addition to establishing a rail system on our interstate highways, we should restore the railroads that once linked our small towns and villages to cities. In most cases, this will be as simple as laying new rails on the old grades, installing crossing gates, and rebuilding decayed bridges. These small rural trains should be battery powered and operated robotically, leaving the conductor free to collect fares and provide comfort and security.

Besides replacing much of our car and truck traffic, efficient trains can replace much of our aviation transportation. Aviation accounts for about 2.5 percent of total carbon emissions. Airplanes are inherently inefficient, because much of their fuel is spent to counter gravity, leaving only the remainder for propulsion. Another problem with air transport is that our current network of airport hubs often requires that passengers fly in the wrong direction before they can fly in the right direction. This is doubly wasteful, for it not only adds hundreds of miles to flights (as well as time and stress) but requires two ascents to flight altitude during which the engines consume fuel at the maximum rate. Given the inefficiency of air transport, governments should stop subsidizing it. Instead, the full costs should be borne by the passengers and the airfreight companies. These costs would include air traffic control, airport security, and airport construction and maintenance. This would make air travel much more expensive and cause many more people to switch to high-speed trains for domestic travel and ships for overseas travel. For journeys of 300 miles or less, trains would actually be faster than planes, because there would be no need to spend time driving to the airport, finding parking, taking the shuttle bus, checking in, passing through security, and trekking to the proper gate. There would also be no more delays due to bad weather.[2] An alternative option for short flights could be airships (or airship-airplane hybrids). Airships require no energy to stay aloft and are therefore extremely fuel efficient. (The Graf Zeppelin circumnavigated the globe in 1929 with only three stops to refuel.)

For transportation within cities, electric streetcar systems should be restored. Compared to diesel buses, electric streetcars are low polluting, require much less money to power and maintain, and last at least three times longer. (They don't, for example, have rubber tires that need to be periodically replaced and disposed of.) Nowadays, streetcars could even be operated robotically, leaving the conductor free to collect fares and ensure passenger safety. They could also be battery powered, thereby eliminating the need for unsightly and expensive overhead wires and trolleys. America once had outstanding urban streetcar systems, but from the late 1930s to the early 1950s, three holding companies controlled by the General Motors Corporation purchased about 150 of America's streetcar systems and converted them to diesel bus systems.[3] These new bus companies were bound by contracts that required them to purchase only GM and Mack buses, use only Firestone tires, and consume only Standard Oil diesel fuel. The three companies involved were eventually tried and convicted of violating antitrust law, but by then America's streetcar systems had been destroyed. As a nation, we are still paying dearly for this betrayal of the public trust. What was good for GM was deeply harmful to America and (as we now realize) harmful to the world's climate.

Commuter ride-sharing and hitchhiking are other efficient ways to transport people. Fifty years ago, the Canadian government under Prime Minister Pierre Trudeau funded a system of summer youth hostels in towns and cities across Canada. The idea was to make it affordable and safe for young Canadians to hitchhike around Canada, so they could see their country and get to know Canadians from other provinces. These hostels were usually located in rented houses and were operated by young people. The cost of a night's stay was about 50 cents. The Canadian government subsidized this hostel system until 1976. During its heyday, it enabled hundreds of thousands of young people to travel many millions of miles without causing significant environmental harm. Unfortunately, hitchhiking has fallen out of favor in recent decades, as people have lost trust in one another. But in a less populated society, with reduced economic and social divisions, trust will be restored, and people will once again gladly pick up hitchhikers and hitchhike themselves. This would be a friendlier and cheaper alternative to Uber or Lyft.

We should also strive to make our communities safer for pedestrians and cyclists. Whenever feasible, children should walk or bicycle to school (and we should once again have neighborhood schools). This will reduce both pollution and obesity and give parents more free time.

Most of the world's people dream of owning a car. They don't realize that cars are not glamorous luxuries providing freedom, but burdensome necessities forced upon us by lack of good public transportation. We end up working much of our lives to pay for car loans, car insurance, license fees, gasoline, tolls, parking, repairs, and taxes for road construction and maintenance. We also have to pay the costs of oil spills, as well as the health costs of air pollution and collision injuries. And now we face the massive costs of climate change, as oceans rise and climate becomes erratic. We don't ride cars: they ride us. A wise people would resist the siren call of the privately owned automobile and insist on saner alternatives.

Economic globalization (the so-called free trade system) has caused a great increase in petroleum consumption, because much of our food is now shipped long distances. This not only wastes non-renewable fuel, but the resulting pollution contributes to climate change. We can counter this by purchasing as much of our food as possible from local growers at farmers markets and food co-ops or by growing our own food.

Increasingly, the internet is eliminating the need for travel. Businessmen already use teleconferencing extensively, and this has no doubt resulted in better decision-making, since it eliminates jet lag. The COVID-19 pandemic forced millions of office workers to learn to work from home. The ecological benefits of this move included lower carbon emissions, cleaner air, and lower vehicle maintenance costs. COVID-19 also greatly reduced international tourism, which accounts for about 8 percent of global carbon emissions.[4] A secondary benefit of reduced travel is that it reduces the cross-border transmission of harmful microorganisms. We can now take virtual online tours without the expense and trouble of travel. Of course, many people will still want to travel for adventure or to enjoy a warmer climate in winter.

Notes

1. Transportation contributes about 15 percent of all human-generated CO_2.

2. The COVID-19 pandemic has highlighted the need for safer public transportation. All trains and buses need to have ventilation systems with high air exchange rates and high-efficiency filters.

3. Jane Jacobs, *Dark Age Ahead* (New York: Vintage Books, 2004), 39 and 186–88.

4. *Growthbusters* podcast, October 27, 2020.

Part 5:

Political and Economic Reforms

Creating Genuine Democracy

Let every man make known what kind of government would command his respect, and that will be one step toward obtaining it.

—Henry Thoreau, *On the Duty of Civil Disobedience*

Our admirable "Can do!" attitude in the face of technological problems has, curiously, no equivalent in the political realm. Unbounded optimism in the first area is almost universally coupled with complete pessimism in the second.

—Garrett Hardin[1]

The best cure for the ills of democracy is more democracy.

—Edward Abbey[2]

Small populations, no matter their nominal form of government, are inherently democratic, because every citizen enjoys considerable political influence. Large populations, in contrast, are always undemocratic, because each legislator or autocrat represents so many people that the individual has little influence.

A citizen of the Principality of Liechtenstein, whose population numbers less than 14,000, desirous to see his Serene Highness the Prince and Sovereign, Bearer of many exalted orders and Defender of many exalted things, can do so by ringing the bell at his castle gate. However serene his Highness may be, he is never an inaccessible stranger. A citizen of the massive American republic, on the other hand, encounters untold obstacles in a similar enterprise.

—Leopold Kohr[3]

Although reducing population size would go a long way toward improving democracy, other measures also need to be taken. What follows are a few practical steps that Americans could take to revitalize their faltering democracy.

Public Financing of Political Campaigns

Currently, incumbent politicians enjoy a strong electoral advantage because they can easily raise more money for an election campaign than their opponents. The donors of this money expect a good return on their investment, and they are rarely disappointed. If we want our politicians to represent our interests, we will have to publicly fund their campaigns. There is no alternative. Public campaign financing is the single most important reform we could make to improve democracy in the United States.

I envision public financing of major offices this way:

- Ban all paid election advertising during the 30-day period preceding an election (citizens could, of course, still place signs in their front yards or sport bumper stickers or speak out in favor of their favorite candidates).
- Require all candidates to completely fill out a detailed questionnaire that would disclose their views and positions (whether genuinely held or not) on a wide range of relevant topics.
- Require all candidates to appear together in a series of public forums during the 30 days preceding the election.

Half of the questions in the candidate questionnaire could be chosen by internet voting and the other half by a committee selected by the 2,100 members of the National Academy of Sciences. Candidates would be required to answer every question in the questionnaire. They would be granted a limited number of words for each answer.

Six forums in which all the candidates would appear together would be scheduled during the 30-day campaign season. These forums would be televised by all television stations within the candidate's district, aired on all radio stations, and available for viewing on the internet. In each forum, each candidate would be allowed 10 minutes to present his or her ideas. When the candidates were finished with their presentations, they would each have another 10 minutes to critique the views of their opponents. If we limit the number of candidates to six, each forum would last two hours. Any candidate who failed to show up for a forum would have his or her name removed from the ballot.

To qualify for public funding, candidates would need to gather as many signatures as possible from registered voters. If the maximum number of candidates was six, then the top six signature gatherers would stand for election. Only unpaid volunteers could gather these signatures. Voters would not be allowed to sign more than one candidate's petition for a particular office or we might end up with six candidates who all represented the views of whatever political persuasion was numerically dominant. The names of the top six in gathering signatures would be announced 30 days before the election. All six candidates would get exactly the same amount of public money to fund their campaigns, and they could spend no more—not even their own money.

The amount of public funding received by each candidate would suffice to pay for a rental car, motels, meals, telephone bills, and the costs of operating a website during the 30-day campaign season. This would cost very little. For a presidential election, the cost could be as low as one penny per citizen, which would amount to about $3 million. If that were the case, each of the six candidates would receive $500,000. That's more than enough for 30 days of motel rooms, meals, rental cars, telephone bills, staffing, and running the website. No donations from citizens, unions, or corporations would be permitted, although anyone could voluntarily canvass for their preferred candidate.

Citizens would have two reliable sources of information to guide their voting: the answers to the campaign questionnaires and the presentations of the candidates during the forums. During the 30-day campaign season, citizens would no longer be bombarded by paid political ads full of deceit, lies, and demagoguery. Without these paid political ads, it would be much harder to bamboozle and frighten the voters. We should also enact stronger libel laws, so that the internet companies that provide venues for libel can be prosecuted.

Knowing the actual (or claimed) views of candidates would enable us to vote for the candidate who best represents our own ideas about best governance. No matter which candidate was elected, he or she would no longer be beholden to wealthy interests.

Those aspiring to become candidates would be free to spend money to promote themselves *prior* to submitting the signatures gathered in support of their candidacy, but they would be required to reveal the names of all campaign donors. The focus of such spending would be to generate excitement about the candidate in order to inspire volunteers to go out and collect enough signatures to get him or her qualified to appear on the ballot. Even so, with up to six candidates on the ballot, and all six receiving exactly the same amount of money with which to run their campaigns during the 30-days prior to the election, there would be fair competition in our electoral contests.

Once public campaign financing is adopted, all existing laws mandating term limits for political office should be abolished, including the 22nd Amendment of the US Constitution (which limits US presidents to two terms). These laws were enacted to try to counter the advantages incumbents enjoy. But they failed to do so, because they failed to regulate the flow of money that confers those advantages. Term limit laws are condescending to voters and inherently undemocratic. In a democracy, voters must be free to choose the leaders they want, even if they want the same ones over and over.

Removing special interest money from the election process would *not* constitute denial of free speech (notwithstanding the US Supreme Court's *Citizens United* decision), because money given in return for favors is neither speech nor a true gift. By effectively giving large donors the power to write the nation's laws, the Supreme Court has dangerously undercut the ability of ordinary people to influence government policy.

Because public campaign financing would lead to a much more representative Congress, it might lead to the impeachment and removal of Supreme Court justices who oppose genuine democracy. Once those justices are removed, the Court could begin to represent the interests of *all* the people, not just the corporate elite and the parasitic financial sector.

A major benefit of reducing the money threshold required to run for office is that many more women would become candidates, and many more would be elected. As of 2021, women comprised only 24 percent of the US Senate and 27 percent of the House of Representatives. These extremely low percentages are not the result of a strong preference on the part of voters for male leadership, but a consequence of so few women appearing on the ballot. As a result, we do not have a government "of the people, by the people." A government with many more women in it would likely be wiser, less aggressive, less venal, and more truthful.

In the aftermath of the corruption and near despotism of the Nixon administration, Senator George McGovern expressed optimism that campaign reform was imminent:

> Today the prospects for further restrictions on private campaign financing, full disclosure of the personal financing of the candidates, and public finance of all federal campaigns seem to me better than ever...[4]

McGovern was right: the prospects were much better. Unfortunately, not enough politicians were willing to act and the people failed to make them act.

Proportional Representation

In earlier editions of this book, this section was called "Eliminating Gerrymandering." In it, I made the point that gerrymandering is damaging to democracy and I recommended a fair way to draw congressional districts to make them competitive. But after reading an article by Lee Drutman,[5] I came to understand that if all congressional districts were competitive, this would create another set of problems. For example, the slightest change in public opinion from one election to the next could produce radical swings in which party prevailed. The members of the losing party might end up with *no* congressional representation. Another problem is that Democrats predominantly live in cities while Republicans mostly live in rural areas. That makes drawing geographically continuous districts that are competitive very difficult.

Drutman proposed that instead of working to end gerrymandering, we should adopt a system of proportional representation—a system that would ensure that *every* vote counts. This could be accomplished by holding statewide elections for congressional representatives rather than district elections. The state's congressional seats would then be distributed proportionally among the political parties. If a state was allocated 8 representatives and the vote was 50 percent for Republican candidates and 50 percent for Democrats, 4 seats would be assigned to the top vote getters from each party. And if the vote was 50 percent for Republicans, 25 percent for Democrats, and

25 percent for third-party candidates, the Republicans would get 4 seats, while Democrats and third-party candidates would each get 2 seats. No longer would third-party candidates function as "spoilers." And states that are solidly Republican or Democratic would finally get a few representatives from the other parties. By enabling each vote to count, we would reduce the apathy and anger that inevitably arise when people feel unrepresented by their government.

The House of Representatives is not the only branch of Congress that could be made more representative. We could make the votes of senators representative too (as James Madison urged). For example, we could have the votes of each of the two senators from California (the most populous state) count as 1 (as they now do), while the votes of senators from other states would count a fraction of 1 depending on their population size relative to California's. Thus, the senators from America's least populous state, Wyoming, would have votes that count only 0.015 compared to California's 1.0. That would be fair, since California's population is 691 times greater than Wyoming's. States that oppose the idea of proportional senatorial votes would be free to secede. One advantage of a more representative Congress is that it would make it much easier to amend the Constitution so that it could reflect contemporary values.

Another way to create proportional representation would be to institute direct democracy at the federal level. Some progressive states and cities already allow their citizens to hold referenda on controversial issues, provided they can collect enough signatures on petitions. This has enabled some states to eliminate laws that criminalize possession of cannabis. It's high time to adopt the referendum process at the federal level, so that the will of the people can prevail over venal and out-of-touch federal legislators. A referendum on abortion rights, for example, would easily pass in the United States. There is already a movement in the US to establish a national referendum process. It is called the National Citizens Initiative for Democracy.

Full Voting Rights for the Residents of the District of Columbia and US Territories

A long overdue reform in the United States would be the granting of congressional representation to the residents of the District of Columbia. These citizens pay federal taxes and have been conscripted to fight wars, yet they have no voting representative in either the House of Representatives or the Senate. Because the Constitution delegates congressional representation to the states, a simple and practical solution would be to return the District of Columbia to Maryland. This would give the residents of DC representation in both the House of Representatives and the Senate. It would also permit DC to have a city government free from congressional meddling.[6] Only actual federal properties would remain under federal control. Currently, the state of Maryland does not want DC returned, but Congress could placate Maryland by providing long-term funding to cope with DC's large social and economic problems. No reasonable person would object to this solution. The denial of congressional representation to the citizens of the District of Columbia is a clear violation of the 14th Amendment's due process and equal protection clauses. The US Constitution contradicts itself: on the one hand, it leaves the citizens of the District of Columbia without congressional voting rights; on the other hand, it demands due process for all citizens.

Voting rights should also be extended to the residents of all US territories (Puerto Rico, the US Virgin Islands, Guam, American Samoa, and the Northern Mariana Islands). Like the residents of DC, these US citizens have no federal voting rights, yet they are required to register for military conscription if they are male and less than 26 years old.

Constant vigilance is necessary against efforts to deny voting rights. Many states currently deny voting rights to those who are in prison or have previously served time in prison. Such practices violate both the 14th and 15th Amendments of the US Constitution and should be abolished by the Supreme Court. Nor should anyone have to register before voting. There is no need for it. A driver's license or state-issued identification card is sufficient to establish one's identity and place of residence. Nothing else is needed to establish whether one is qualified to vote in a particular district. Computer databases could easily be set up to prevent a person from voting twice. Mail-in voting should also be an option available to everyone. Laws that make it more difficult to vote have only one purpose: to subvert democracy. The people behind these laws need to be removed from office before we lose the power to remove them.

Elimination of the Electoral College

The Electoral College was incorporated into the Constitution as a last-minute compromise between those who wanted to elect the president by a vote of Congress and those who wanted to elect the president by a popular vote of enfranchised citizens. Poll after poll has demonstrated that most Americans want the Electoral College abolished, yet Congress has steadfastly ignored the people's will.

On five occasions, the Electoral College installed a president who lost the popular vote: in 1824, Andrew Jackson won the popular election, but the Electoral College gave the presidency to John Quincy Adams; in 1876, Samuel Tilden won the election, but the Electoral College gave the presidency to Rutherford Hayes; in 1888, Grover Cleveland won the election, but the Electoral College gave the presidency to Benjamin Harrison; in 2000, Al Gore won the election, but the Electoral College (abetted by a deeply biased right-wing Supreme Court) gave the presidency to George W. Bush; and in 2016, Hillary Clinton won the election by some 3 million votes, but the Electoral College gave the presidency to Donald Trump.

At the US Constitutional Convention held in 1787, delegates from the slaveholding states argued that for purposes of determining congressional representation, slaves should be considered full persons, since that would give slave states more representatives and more power (slaves, of course, could not vote). But when it came to levying federal taxes on the states, the slaveholding states took the opposite stance, arguing that slaves should be considered property because that would reduce the state's tax burden (federal taxes were based on population, not property value). As a compromise, the delegates at the Constitutional Convention decided that slaves would constitute 3/5 of a person for congressional representation and federal taxation. That gave Southern states 1/3 more representatives and 1/3 more electoral votes than they would have had if slaves had been classified as property. Not surprisingly, 10 of the first 15 US presidents were slaveholders from Southern states. The Southern states continued to wield undue power over presidential elections until the passage of the Voting Rights Act in 1965, which guaranteed all citizens the right to vote. Although the 3/5 compromise (and subsequent Jim Crow laws that disenfranchised blacks) no longer benefit Southern states, the Electoral College has not been abolished. As a result, the only states that matter in presidential elections are the so-called battleground states where the election outcome is uncertain. Citizens in all other states (including all states with small populations) are completely ignored in presidential campaigns. That gives voters in battleground states a disproportionate and unfair influence over the outcome of presidential elections. This is because state electors are chosen based on which candidate wins the state popular vote, not on which candidate wins the national popular vote.

One way to eliminate the Electoral College would be to amend the Constitution. This, however, would be a long and difficult process. A simpler approach would be to leave the Electoral College in place but require that each state cast its electoral votes for whoever wins the national popular vote (instead of casting them for the winner of the state popular vote). A bill to do this, called the National Popular Vote Bill, has already been passed by 15 of the 50 state legislatures and the District of Columbia (see www. nationalpopularvote. org). These states represent 195 electoral votes. A further 75 votes are all that is needed to cross the threshold of 270 electoral votes, whereupon the bill would become federal law. If this bill should ultimately fail, there is something else that could be done: the two leading presidential candidates could agree with a handshake to respect the popular vote. Whichever candidate received fewer popular votes would withdraw his candidacy (and that of his vice-presidential running mate) just before the votes of the electors were announced. Perhaps this would precipitate a constitutional crisis, but so what? A crisis may be what is needed to finally motivate Congress to abolish the antidemocratic Electoral College.

Of the five reforms proposed above, campaign finance reform is by far the most important. This is because its enactment would produce a more representative Congress, which would then be more likely to enact the other reforms. However, the present Congress will not carry out any of these reforms unless the public demands them, and that won't happen until the public is educated enough to understand the gravity of the internal threats democracy now faces. Jefferson gave us fair warning: "If a nation expects to be ignorant and free, in a state of civilization, it expects what never was and never will be."[7]

Forty years ago, Edward Abbey summed up our political predicament:

> We call our system a "representative democracy" but in fact our representatives, with honorable exceptions here and there, represent not the voters, but those who finance their election campaigns. In the Soviet Union, the egalitarian ideal of theoretical communism was betrayed from the beginning; in the United States, the Jeffersonian vision of a decentralized society of independent agrarian freeholders was dead by the end of the nineteenth century while democracy, defined in Lincoln's words as "government by the people" has never even been tried.[8]

Notes

1. Garrett Hardin, *Filters against Folly* (New York: Penguin Books, 1985), 213. Copyright © 1985 by Garrett Hardin. Used by permission of Viking Books, an imprint of Penguin Publishing Group, a division of Penguin Random House LLC. All rights reserved.

2. Edward Abbey, *The Journey Home* (New York: Penguin Books, 1991), 230. Copyright © 1977 by Edward Abbey. Used by permission of Dutton, an imprint of Penguin Publishing Group, a division of Penguin Random House LLC. All rights reserved.

3. Leopold Kohr, *The Breakdown of Nations* (Devon: Green Books Ltd., 1957, 2012), 113.

4. *Washington Post*, August 12, 1973.

5. Lee Drutman, "A Better Way to Vote," *Time*, February 12, 2022, pp. 31–32.

6. In 2011, for example, Congress forbade the DC government from using its own revenues to pay for abortions for women on Medicaid.

7. Thomas Jefferson in a letter to Charles Yancey, January 6, 1816. See www.monticello.org.

8. Edward Abbey, "A Writer's Credo" in *One Life at a Time, Please* (New York: Henry Holt and Company, 1977-1988), 169.

Reforming the Trade System

The current global free trade system causes harm in several ways:

1. Free trade allows corporations to externalize costs by manufacturing in countries where pollution is poorly regulated. This contaminates poor countries and undercuts industries in more advanced countries where pollution is better regulated.

2. Free trade allows overpopulated countries with low wages to attract industries from less populated countries with higher wages. This rewards overpopulated countries while harming countries that have had the good sense to restrain their fertility.

3. Free trade allows the owners of foreign companies to control the economic fate of our communities—communities about which they know little and care nothing.

4. Free trade drives many countries into deep debt from which they try to escape by overexploiting natural resources and inflating their currency (which allows them to pay back their debts with currency of lower value). But inflation causes economic hardship for the population as the purchasing power of salaries and pensions

shrinks. This, in turn, can lead to political instability, capital flight, and ruinous currency speculation.

5. Free trade works against sustainable economies because the entire system of free global trade is predicated on massive resource consumption and massive pollution. There is nothing sustainable about it. A saner society would reject free trade in favor of protecting its own people and industries. We should never allow ourselves to become vulnerable to trade disruptions or blackmail by other countries. This is a matter of national security, as well as a matter of personal well-being and cultural sustainability.[1]

To properly regulate free trade and eliminate the trade deficit, we should insist that imports balance exports. This would automatically prevent free capital mobility and preclude the accumulation of a large national debt. (Nations get into debt mainly because they import more than they export.) A logical way to make imports balance exports would be to estimate the value of what the country will export during the coming year, and then issue import permits equivalent to that amount. These import permits would be auctioned off to import firms and would be valid for one year. If an import firm ended up not using all of its permits, it could sell them to other firms eager to import more.[2]

Notes

1. For a good discussion of the harm of free trade see *Beyond Growth: The Economics of Sustainable Development* by Herman Daly, 160–166.

2. The idea of using import permits to eliminate trade deficits was proposed by Herman Daly and John B. Cobb in their book *For the Common Good: Redirecting the Economy Toward Community, the Environment, and a Sustainable Future*, p. 230.

Reforming Charity

When goodness is lost, philanthropy appears.
—Tao Te Ching[1]

Before undertaking any efforts to assist others through charity, we should carefully weigh the words of Thoreau:

> Be sure that you give the poor the aid they most need, though it be your example which leaves them far behind. If you give money, spend yourself with it, and do not merely abandon it to them.[2]

And this:

> There are a thousand hacking at the branches of evil to one who is striking at the root, and it may be that he who bestows the largest amount of time and money on the needy is doing the most by his mode of life to produce that misery which he strives in vain to relieve.[3]

When charity is used to reduce mortality without simultaneously reducing the birth rate, the result is population increase. This obliges breadwinners to support more children and more elderly relatives, resulting in more forests converted to agriculture, shorter fallow periods, overgrazed pastures, overfished lakes and seas, and conflicts with neighbors over vital resources. At some point, the population will have to stop growing, but it would be far better for all if the birth rate were lowered at the same time as the death rate. That way environmental ruin and suffering could be avoided.

The world's largest private charity is the Bill & Melinda Gates Foundation. In 2012, the Gates Foundation announced that it would begin a major effort to make contraception available at affordable prices to more of the world's women. It also announced it would support research to develop better contraceptives. This is commendable and represents a major shift for the Gates Foundation, which had previously focused exclusively on projects aimed at reducing premature mortality.

The Gates Foundation made an egregiously bad grant in 2008 when it gave $42 million to Heifer International to fund dairy enterprises in Africa. Heifer International is an organization that distributes domestic animals to poor people around the world. This animal trafficking is supposed to benefit the poor, but the real beneficiaries are the animal breeders and the charity's own bureaucracy. Here is what Heifer International proposed to do in East Africa:

> It almost sounds like a joke. Set up dairy enterprises in rural African villages with no refrigeration, electricity, veterinary care or passable roads for a population that can't drink milk because it's 90% lactose intolerant. But the Bill & Melinda Gates Foundation didn't think it was a joke when it announced

the gift of $42 million to Heifer International at the World Economic Forum in Davos, Switzerland in January—the biggest gift the Little Rock, AR-based Christian charity which sends live animals to poor countries has ever received...To get around the lack of rural electricity for the proposed dairy operations in Kenya, Uganda and Rwanda, Heifer will create "chilling plants" with their own backup power generators according to a press release where the milk will be stored for pickup by "refrigerated commercial dairy delivery trucks"—both of them. Farmers will artificially inseminate cows, perhaps by candlelight, with "high-production dairy animal semen"—more backup generators required to keep it frozen?—and increase milk quality through providing "improved animal nutrition" to the cows with the food they don't have.[4]

Poverty can be eliminated, but not by philanthropists. It will be eliminated when there are no more philanthropists.

> The sage does not hoard, and thereby bestows.
> —*Tao Te Ching*[5]

Notes

1. Lao Tzu. *The Tao Te Ching of Lao Tzu*, trans. Brian Browne Walker (New York: St. Martin's Griffin, 1995), chapter 38.

2. Henry David Thoreau, *Walden* (1854), "Economy" chapter.

3. Ibid.

4. Martha Rosenberg, "Why Heifer International Is Rolling in Dung," *Counterpunch*, July 12, 2008. Reprinted with the permission of the author.

5. Lao Tzu. *Tao Te Ching: A New Translation*, trans. Sam Hamill (Boston: Shambhala Publications, Inc., 2007), chapter 81.

Eliminating Poverty

Heaven's way reduces surplus and supplements insufficiency.

—*Tao Te Ching*[1]

It is a subject often started in conversation, and mentioned always as a matter of great surprise, that, notwithstanding the immense sum which is annually collected for the poor in this country, there is still so much distress among them.

—Thomas Malthus[2]

Poverty is the want of basic shelter, clothing, food, fuel (for cooking and heating), medicine, and physical security. It generates a witches' brew of social ills. Some societies foster poverty; others eliminate it.

Poverty is increasing in the United States. Beginning around 2000, employment and median income began declining even as GDP and corporate profits rose. Today, the US middle class is smaller than the poor and the rich combined, and the income gap between the rich and everyone else is expanding rapidly. According to economist Joseph Stiglitz, the richest 1 percent in America now captures nearly a quarter of the nation's income every year and controls 40 percent of the nation's wealth. The richest 1 percent has also captured most of the income growth in the years since the 2008 economic crash—an amazing 73 percent.[3] The popular cheer to "grow the economy" will not save the 99 percent. Even if growth were sustainable (which it obviously is not), the benefits would go largely to the rich.

Jobs that pay a living wage are disappearing because of mass immigration, laborsaving technology, and the offshoring of jobs to countries where wages are low and regulations few. Mass immigration, by flooding the labor market, contributes to the weakening of labor unions and the displacement of native workers—especially black workers. The disappearance of good jobs forces people to fall back on the charity of relatives or rely on part-time work at minimum wage without benefits. About half of all US jobs are now low wage, and the Bureau of Labor Statistics forecasts that two-thirds of *new* jobs will also be low wage.[4] A study by the McKinsey Global Institute concluded that 45 percent of all the activities that American workers engage in today could be automated with *existing* technology.[5] If computers get just a little better at natural language (as they surely will), a depressing 58 percent of worker activities could be automated. The upcoming job losses will impact workers at all income levels—blue collar and white collar alike. The crux of the problem is this: the unspoken goal of today's technological innovators is an unemployment rate of 100 percent. Ponder that.

The primary beneficiary of labor elimination is employers. Not only do they profit from paying fewer salaries, but high unemployment means they can pay less to their existing workers (who are desperate to keep their jobs). And there is no danger that these workers will slack off or grow restive and demand more pay.

Poorly designed international trade agreements are another major cause of unemployment. These agreements give corporations a strong economic incentive to produce their products in poor countries where labor costs are low and environmental regulations are few or nonexistent. This has brought about the deindustrialization of much of the United States and the loss of millions of well-paid American jobs. Today, not only production but services as well are being offshored at an accelerating pace.

When people are pushed out of full-time employment, they lose not only a reliable and adequate income, but are deprived of employer-paid medical insurance, as well as paid vacation and a retirement program (pension or 401[k]). Job loss also damages self-esteem and triggers depression which can lead to addiction, domestic violence, and permanently scarred children.

In the face of the ongoing loss of jobs that pay a living wage, several stopgap measures have been proposed. These include raising the minimum wage, reducing the length of the work week (so that more people can be employed), making it easier for workers to unionize, renegotiating international trade agreements (to preserve more jobs), reforming the miserly Earned Income Tax Credit (EITC),[6] and having the federal government invest massively in repairing infrastructure in order to increase employment and preserve the infrastructure that the economy depends upon. These well-intentioned proposals can certainly help in the short run, but in the long run they cannot cope with the looming crisis. Moreover, there are problems with some of these proposed measures. For example, raising the minimum wage, shortening the work week, and unionization all depend on the existence of traditional full-time jobs and a tight labor market. These measures would do little to help the tens of millions who have already been pushed into the world of contingent ("gig") labor.

And shortening the work week would require significantly reducing the pay of those who were required to work fewer hours.

Among those who have seriously considered the impending unemployment crisis, there is a growing consensus that two steps must be taken: put an end to mass immigration and implement a universal basic income (UBI)—an income to which all citizens would be entitled as their birthright. A UBI would close the gap between the rich and the rest of society by equitably distributing the immense profits generated by machines. It would also cause money to circulate freely throughout society instead of being concentrated at the top. This would alleviate economic distress and prevent violent social upheavals.

To permanently eliminate poverty, a UBI would need to be combined with a system to prevent an increase in the birth rate. Otherwise, the extra income provided by a UBI would tempt many people to have more children, which would undercut the twin goals of eliminating poverty and reducing population size.

The concept of the UBI has a long history. The revolutionary writer Thomas Paine proposed it in 1795 in his book *Agrarian Reform*. The idea was resurrected in 1946 by economist George Stigler, who called it a "negative income tax." In the 1960s, the idea gained more traction, and, during the Nixon administration, a law to establish a "guaranteed minimum income" came within a hair's breadth of being enacted by Congress. (The House of Representatives passed it, but the Senate balked.) Among the prominent people who have endorsed one version or another of a UBI are conservative economist Milton Friedman, Nobel economist F. A. Hayek, US President Richard Nixon, civil rights leader Martin Luther King Jr., economist Herman Daly, former union leader Andy Stern, Carl Camden (CEO of Kelly Services), Ralph Nader (consumer advocate and presidential candidate), Vinod Khosla (Sun Microsystems founder), Amitai Etzioni (George Washington University sociologist), Charles Murray (libertarian political scientist), and Robert Reich (former US Labor Secretary). A UBI would eliminate the entire chaotic, ill-conceived, inefficient hodgepodge of existing welfare programs and replace it with the only true antidote to poverty: money.

Social experiments have already shown that a UBI would work well. Alaska, in the 1970s, established the Alaska Permanent Fund, which distributes money each year to all Alaskans. This money comes from payments that oil companies make to the state for oil leases. Currently, the citizens of Alaska receive about $2,000 each year. This money has substantially alleviated poverty in Alaska. Another UBI experiment was conducted by the Eastern Band of Cherokee Indians in North Carolina. They decided to share the proceeds from their tribal gambling casino with all the members of the tribe. Again, the result was the alleviation of poverty and stress.

What follows are my own thoughts on how a universal basic income could be established in the United States. As previously mentioned, such a system would have to be combined with an effective system to regulate birth rates and immigration. Otherwise, the population would continue to grow, and this would drive down wages by swelling the labor supply.

In my plan, every citizen—whether rich or poor—from the age of 18 onward would receive an annual UBI of $20,000 (enough to survive on in 2023). The payments would begin each April 15 (the traditional US tax deadline). They would be made biweekly to a debit card or bank account. Citizens could also take their income as a lump sum (perhaps to start a business or purchase a house).

After one year, on April 15, those whose incomes exceeded $75,000 during the preceding year would have to return the entire $20,000. For them, the money would have served as an interest-free $20,000 one-year loan. Those who had no income during the year would keep the entire $20,000. People whose incomes ranged between $1 and $75,000 would return a portion of the UBI equal to 25 percent of their earned income. So someone who earned $20,000 would return $5,000 and keep the remaining $15,000, giving them a total income of $35,000. The more people earn, the more they would have to return, as shown in the following table:

Earned Income	UBI Amount ($20,000 – 1/4 of Earned Income)	Total Income (Earned Income + UBI)
$0	$20,000	**$20,000**
$5,000	$18,750	**$23,750**
$10,000	$17,500	**$27,500**
$15,000	$16,250	**$31,250**
$20,000	$15,000	**$35,000**
$25,000	$13,750	**$38,750**
$30,000	$12,500	**$42,500**
$35,000	$11,250	**$46,250**
$40,000	$10,000	**$50,000**
$45,000	$8,750	**$53,750**
$50,000	$7,500	**$57,500**
$55,000	$6,250	**$61,250**
$60,000	$5,000	**$65,000**
$65,000	$3,750	**$68,750**
$70,000	$2,500	**$72,500**
$75,000	$0	**$75,000**

Anyone who failed to return the owed portion of their income would have that amount deducted from the following year's UBI. So if someone earned $40,000 and failed to return the $10,000 that was due, they would only receive $10,000 the following year, instead of $20,000.

Consider a household made up of a father, a mother, and two grown children—all of them unemployed. Under this proposal, that family would now have a combined household income of $80,000. That means one of them could attend college, or they could establish a small business. And the money they spent would stimulate the local economy.

Most Americans would strongly support a UBI program, because it would benefit them materially. The question then becomes, "Can the United States afford such a program?" To answer that, we first need to figure out what the program would cost. To do that, we have to find out how many Americans over the age of 18 earn nothing annually, how many earn more than $75,000, and how many are in the various categories in between. Then we can set up a table in which we use the mean income within each income bracket for our calculations. For this exercise, the deduction from the UBI for those who earn between $1 and $75,000 will be set at 25 percent. People who earn nothing at all (the unemployed and stay-at-home parents) would receive the baseline income of $20,000.

Annual Income	Amount of UBI Retained (See note 1 below.)	No. of Adults (in millions) in Each Income Category (See note 2 below.)	Cost of UBI (in billions) at End of 1st Year (column 2 x column 3)
$0	$20,000	100.15 (See note 3 below.)	$2,003
$1-$10,000	$18,750	22.053	$413.50
$10,000 to $20,000	$16,250	23.621	$383.80
$20,000 to $30,000	$13,750	18.881	$259.60
$30,000 to $40,000	$11,250	14.600	$164.30
$40,000 to $50,000	$8,750	11.473	$100.40
$50,000 to $75,000	$4,375	19.395	$84.90
>$75,000	$0	36.551	$0
		Total first-year cost:	$3,409.5 billion ($3.409.5 trillion)

1. This is ¼ of income subtracted from UBI for categories between $1 and $75,000.

2. 2014 data from Internal Revenue Service tax table: https://www.irs.gov/uac/soi-tax-stats-individual-statistical-tables-by-size-of-adjusted-gross-income. Within each income bracket, the mean is used for calculations. Thus, within the $20,000-$30,000 bracket, $25,000 is used to determine the amount of UBI retained.

3. This number represents the 98.116 million adults who did not file tax returns (and thus presumably had no income in 2014) and the 2.034 million who filed but whose adjusted gross income was $0.

The following figures are based on 2014 data, since that was the latest complete tax information available at the time I wrote this chapter. In 2014, there were 320 million people in the United States. Of these, 73.28 million were under the age of 18 and would not qualify for a UBI. So the number that would qualify for the UBI is 246.72 million (320 million - 73.28). From these we must subtract the 36.55 million who earned more than $75,000, because they would have to return their entire UBI after one year. So 246.72 – 36.55 = 210.17 million adults who would qualify to keep all or part of the UBI. Of these 210.17 million, 110.02 million filed tax returns and had an adjusted gross income of $1 or more, while 100.15 million either did not file tax returns or their adjusted gross income was zero. For purposes of calculation, we will assume that those who did not file tax returns had no income at all. Each of these 100.15 million people would keep the entire $20,000 UBI, so the total cost for their UBIs would be $2.003 trillion. To this cost, we must add the cost of those who earned between $1 and $75,000. When we do this (see above table), we find that the total first-year cost of the UBI program would be $3.41 trillion. That is about three times the current (2021) discretionary budget of the US (which is $1.485 trillion).

Although $3.41 trillion is a huge sum, it is nevertheless affordable. In order to pay for the UBI, the government would need to raise three to four times more tax revenue than it currently does. The first step should be to impose a cap on the amount of income people can keep. I would set the cap at 12.5 times the base income of $20,000. So all income over $250,000 would go into the federal coffers. Right now, about 19 percent of total US income goes to just 1 percent of the population (or, more correctly, 1 percent of tax filers). Since the total US income is $16 trillion, 19 percent of this would be $3.04 trillion. That alone would cover nearly all the first-year cost of the UBI ($3.41 trillion). The remainder of the cost, as well as what is needed for the rest of the discretionary budget, can be obtained from those who are slightly lower on the wealth scale than the top 1 percent. And still more revenue could be raised by closing tax loopholes, catching tax cheats, and imposing heavy inheritance taxes (to prevent unearned wealth from creating greater inequality). And keep in mind that a great deal of the money which is currently spent on poverty programs, such as the $75 billion spent each year on the SNAP (food stamp) program, would become available to help fund the UBI. Reforming the American diet is another way we could free up an immense amount of money to pay for the UBI. Currently, the US spends $650 billion annually on Medicare and $550 billion on Medicaid—that is, $1.2 trillion. Keeping people healthy would eliminate most of these costs. Cutting military spending (which was $767 billion in 2020) is another way to free up money for a UBI. Much of the money spent on the military does not truly contribute to national security. Instead, it enriches defense contractors.

One way to counter tax-cheating corporations (which transfer their capital to countries with near-zero tax rates) would be to tax them according to how much they make in every country where they do business. So, if 10 percent of Apple's profits were made in the United Kingdom, Apple would have to pay taxes to the Exchequer based on that 10 percent. In this way, each country would get its fair share of taxes from Apple's profits. And this would deprive corporations of the incentive to move their capital to other countries.

Implementation of the UBI would eliminate two closely related evils: poverty and extreme income disparity. The rich class would cease to exist. Their great estates will be converted into nature reserves or public parks, their mansions subdivided into apartments or disassembled and recycled, their yachts recycled or turned into research vessels. The rich will, of course, fiercely resist being divested of their wealth, but once the deed is done, they will be much happier, because they will no longer have to worry that someone else has a bigger fortune than their own.

> Superfluous wealth can buy superfluities only. Money is not required to buy one necessary of the soul.
> —Henry Thoreau[7]

The UBI program, due to its permanent stimulation of the economy, would quickly recover much of its own cost. By keeping money in continual circulation in every corner of society, it would increase employment and thereby increase tax revenues. As employment increased, the cost of the program would fall rapidly, because these costs are highest for the unemployed and underemployed.

A UBI would have widespread and deep social benefits. It would bring an end to most street crime, because the destitute (including destitute parolees) would have viable incomes. It would sharply reduce the number of battered women and battered children, because a UBI would give poor women economic independence and reduce the stresses that lead to violence. And it would cause prostitution to decline among the poor. The elimination of begging in the streets would be another welcome benefit, because there would be little sympathy for beggars with a viable income. A basic-income program would also benefit families, since it would allow one spouse to stay at home and care for children and still receive an income, while the other spouse worked outside the home. Alternatively, it would enable them to afford day care. It would also allow people to temporarily quit their job in order to care for an ailing relative (such as a parent with Alzheimer's disease). This would be much less expensive than nursing home care and would be better for the invalid. A basic income would also enable poor women to pay for abortions. This is important, because the US Congress (through the Hyde Amendment) has prohibited the Medicaid program from paying for abortions except in cases of rape or incest. Consequently, a poor woman who wants an abortion must now postpone the procedure until she can scrape together a few hundred dollars to cover the cost. This delay can result in the need for a more expensive procedure with greater health risks for the mother. (If the population-stabilization plan proposed in this book were adopted, abortions would be rare and free, and only the taxpayers who supported abortion rights would pay for the program.) Another benefit of the UBI is that low-income people who have children will no longer have to raise them in poverty. This will have huge benefits for the children and long-term benefits for society. The UBI would cause state and local taxes to fall for everyone, thanks to the elimination of most welfare programs and the elimination of most crime (hence less need for police, courts, and prisons). Further savings would come from not having to spend for private security measures, such as guns, steel doors, barred windows, security cameras, guard dogs, and lights left burning when no one is home. But the greatest benefit of a UBI program would be its permanent stimulation of the economy. By putting money into the hands of people who will promptly spend it to satisfy basic needs, the economy will have a permanent engine to drive it from the bottom up instead of struggling to function on crumbs trickling down from the top. With a basic income program in place, money will circulate properly to all corners of society. Moreover, much of this money will be spent on necessities and services that are produced and provided domestically, thereby increasing domestic employment and reducing the trade deficit.

Another benefit of a UBI is that it would give creative people a viable income, allowing them to develop and express their ideas and talents. They could quit their jobs, or work part-time, and devote themselves to their true calling. Unemployment compensation partially fulfills this role today. The first edition of this book was made possible by unemployment compensation. So were some of Edward Abbey's works. And it was British welfare payments that enabled J. K. Rowling to write the first Harry Potter book (which ultimately brought millions of pounds to the Exchequer). Unemployment compensation has frequently served as venture capital for the poor. A universal basic income will provide much more of this capital—with no strings attached.

An interesting consequence of a UBI program is that the minimum wage could be abolished without harming the working poor. In fact, its abolition would help reduce unemployment, since it would open up additional low-wage jobs. Even if people earned only $5,000 per year, their total income would still be $23,750—well above today's minimum wage income of $15,080. Another interesting consequence of a basic-income program is that it would eliminate much of the need for labor unions. Thanks to the safety net provided by the UBI, workers would feel free to quit their jobs if the pay was too low or the working conditions too harsh and unsafe. And thanks to the high demand for labor created by a shrinking population, they could quickly obtain another job. Consequently, employers would voluntarily offer better pay and better working conditions in order to attract and retain good workers. The result would be a freer, more efficient labor market and less discord between labor and management. Thanks to the income cap, most employers would gain nothing by trying to use the UBI as an excuse to lower wages. Any profits they might make in excess of their cap would go right back to the government, and then back to the people.

The impact of a UBI on reducing racial inequality would be profound. People belonging to disadvantaged racial and ethnic groups would finally be able to buy decent housing and provide their children with a good education. This, in turn, would lead to their children getting higher-paying jobs, helping to break the cycle of poverty. Decades ago, the US federal courts ordered racially segregated US cities to bus black children to schools in white neighborhoods and vice versa. The courts reasoned that this would close the racial education gap. And to some extent, it did. But many white parents who feared crime or were racially prejudiced pulled their children out of public schools and enrolled them in private white schools. This perpetuated educational segregation and weakened public education. School busing failed because it attempted to do an end run around the problem of low income instead of eliminating low income itself.

Those who have paid attention to this proposal will have noticed that it provides a strong incentive for work, because those who work will see their incomes boosted by the basic income (unless they make more than $75,000 annually). Moreover, the people with the strongest incentive to work would be those at the bottom of society, because they will see the greatest supplementation of their earned income by the UBI. So there is no danger that a large proportion of the population would try to live on the basic income alone. Not only would most people work to earn more money, but there would be many more jobs available, thanks to the UBI's permanent stimulation of the economy. And once people have more money in their pockets, they will view the future more positively and be more inclined to work to improve their lot. Malthus understood this:

> The indigence which is hopeless destroys all vigorous exertion, and confines the efforts to what is sufficient for bare existence. It is the hope of bettering our condition, and the fear of want, rather than want itself, that is the best stimulus to industry, and its most constant and best directed efforts will almost invariably be found among a class of people above the wretchedly poor.[8]

A basic-income program would also get more money into the hands of the young, who are more inclined to entrepreneurship than older people, due to their sense of adventure and their willingness to take risks. This would further invigorate the economy. Entrepreneurship could be further encouraged by inviting entrepreneurs to submit business plans and having the government award substantial start-up money to the more promising plans. This approach has worked superbly in a Nigerian experiment, where so many new jobs were created by the entrepreneurs that the money spent by the government was recouped several times over.

To help slow the attrition of jobs, the international free trade system should be reformed. Free trade erodes national sovereignty, aggravates environmental damage, drives down wages, and increases unemployment by forcing American workers to compete against some of the world's poorest people in countries with weak labor and environmental protections. Free trade makes the rich richer at the expense of everyone else and the environment.

With a basic-income program in place, the poor would benefit in multiple ways: they would not be overwhelmed with children they could not afford (thanks to the two-credit system), their earnings would be greatly supplemented by the UBI, and they would have the opportunity to sell one or both population credits for a considerable sum of money. These benefits, combined with a thriving economy and the elimination of mass immigration (resulting in higher wages and less unemployment), would quickly bring an end to poverty.

Any attempt to institute a UBI will be fiercely opposed by those who celebrate selfishness and greed as virtues. They will warn the rest of us against the "moral danger" of giving people something for nothing. Yet these same spokesmen for rugged individualism seem to have no problem accepting inheritances, setting up trust funds for their children, exploiting tax loopholes, receiving government subsidies, and obtaining unearned income as rentiers. It is they—not the UBI—who are the real moral danger.

A basic income would be the birthright of every citizen. It would enable everyone—not just the rich—to live with dignity. It would finally fulfill the promise of "life, liberty, and the pursuit of happiness" for all. Liberated from the necessity to serve a boss or be a boss, people could finally speak their minds. This would increase honesty and improve democracy.

Our dream is to escape the hierarchical order: neither to serve nor to rule.

—Edward Abbey[9]

To protect this social safety net, it is important that the law explicitly recognize that a basic income is an inalienable right of citizenship: a birthright. This will prevent lawmakers from placing
restrictions on the UBI, such as requiring that everyone first pass a
drug test or have no criminal record.

One way people could supplement their UBI would be to acquire a few acres of land for growing food. Anyone familiar with
urban agriculture in Russia knows that city dwellers can raise an
enormous amount of food on absurdly small family plots ("dachas").
Our blighted cities contain thousands of acres of abandoned land
that could be made productive if it were sold cheaply to families with
the will, knowledge, and energy to farm it. Within the city limits of
Detroit alone, there are at least 25 square miles of vacant, abandoned
land (some estimates put the figure as high as 40 square miles). Twenty-five square miles of land could be divided into 16,000 one-acre
allotments (25 sq. mi. × 640 acres per square mile). Some Detroiters
have already begun to farm this vacant land. With proper education
and hard work, urban farmers could be highly productive.[10] A reliable market for their organic produce could be provided if the government purchased the food for local school-lunch programs. Land
whose soil is too polluted for food crops could be planted with trees,
which would capture CO_2 and eventually supply valuable timber.
One of the benefits to these small farmers would be the dignity that
comes from greater economic independence; and this independence
would strengthen democracy, as recognized by Thomas Jefferson:

Corruption of morals in the class of cultivators is a phænomenon of which no age nor nation has furnished an example. It is the mark set on those, who not looking up to heaven, to their own soil and industry, as does the husbandman, for their subsistence, depend for it on the casualties and caprice of customers. Dependance begets subservience and venality, suffocates the germ of virtue, and prepares fit tools for the designs of ambition.[11]

In our own time, writer and farmer Wendell Berry elaborated Jefferson's message:

To speak of the need for settled families and stable communities in rural America is to imply at the same time the necessity of extensive and profound cultural change. Good and responsible use of family-sized holdings cannot be expected of people with the dependent and subservient minds of industrial employees. What is required are people independently intelligent and resourceful, skilled in handwork and practical thought, who have forgotten about "professionalism," "official channels," and "overtime."[12]

We could also reduce unemployment and improve well-being by having the government pay people to do socially valuable work that the private sector does not find profitable. For example, we could hire people to control invasive plant species. These aggressive invaders are self-replicating pollutants that cause enormous economic losses and severe environmental harm.

Other steps we should take to reduce poverty are these:

- Adequately fund public transportation systems so that people can easily get about without the great expense (and great pollution) of a car.
- Establish a vigilant consumer-protection agency with strong enforcement powers.
- Establish a health care system that focuses on prevention.
- Keep Social Security solvent. Social Security functions like a minimum wage for the elderly and disabled. It is an efficiently administered program and nearly fraud-free. It is estimated that it currently keeps about 40 percent of people 65 and older out of poverty.[13] However, keeping the elderly employed longer would also serve to keep them out of poverty, without as much expense to the public treasury.

Eliminating poverty also requires eliminating the crippling debt of individuals and nations. The goal of banks is to maximize profits. They achieve this by maximizing the debt of individuals and nations. The average American now spends 35 to 40 percent of his or her paycheck on debt service. This typically includes a mortgage (average: $168,614), a student loan (average: $48,172), a car loan (average: $27,141), and credit card debt (average: $15,762). Poor nations like Greece likewise spend much of their national income on debt service. The austerity programs that the big banks have imposed on Greece and other poor countries generate high unemployment and poverty, rendering the nation even less capable of paying its debt. The big banks don't care though, because they see this as an opportunity to force privatization of publicly held assets (the nation's commons). Privatization provides the rich with even more opportunities for rent extraction. For example, they can privatize the nation's railroad system, privatize its highways, and privatize its airports and seaports. (A Chinese state-owned company now has a 67 percent stake in the operation of Greece's main seaport.) If the banks were smart, they would moderate their parasitism and let their hosts continue to live. But the global financial system knows no limits and has no prudence. It will bring about its own destruction.

> The financial sector has the same objective as military conquest: to gain control of land and basic infrastructure, and collect tribute. To update von Clausewitz, finance has become war by other means. What formerly took blood and arms is now obtained by debt leverage. Direct ownership is not necessary. If a country's economic surplus can be taken financially, it is not necessary to conquer or even to own its land, natural resources and infrastructure. Debt leverage saves the cost of having to mount an invasion and suffer casualties. Who needs an expensive occupation against unwilling hosts when you can obtain assets willingly by financial means—as long as debt-strapped nations permit bankers and

bond-holders to dictate their laws and control their planning and politics?

—Michael Hudson[14]

Clearly, the long-term economic trend is for robots and computers to permanently displace ever more workers, not only in the manufacturing and service sectors, but in the professions as well. If we are to avoid massive suffering and violent social upheavals, our present economic system will have to be radically overhauled—and soon. We must ensure that the wealth generated by machines (and currently monopolized by a tiny fraction of the population) is equitably distributed. A UBI would accomplish that. Finally, I'll say it once more: it would be folly to implement a UBI without also implementing a program of population regulation to ensure that the combined rates of birth and net migration do not exceed the death rate.

Economist Thomas Piketty pointed out that the reduction of European inequality that took place in the 20th century was due to three factors: the bankruptcies of the Great Depression, the destruction of bond value by inflation, and two world wars. According to him, no society has ever ended its extreme wealth gap by democratic and peaceful means.[15] Perhaps we can be the first. But there is one factor in the reduction of inequality that Piketty ignored: the establishment of strong labor unions. What enabled labor unions to be formed in the US was the curtailment of immigration between 1924 and 1965. This created a tight labor market that obliged employers to accept unions and make wage and benefit concessions. When mass immigration resumed after 1965, labor unions became weak and ineffectual, and that increased the income gap between rich and poor, and especially harmed black people.

John F. Kennedy gave his fellow millionaires fair warning:

If a free society cannot save the many who are poor,
it cannot save the few who are rich.[16]

Notes

1. Lao Tzu, *Tao Te Ching: A New Translation*, trans. Sam Hamill (Boston: Shambhala Publications Inc., 2005), chapter 77.

2. Donald Winch, ed., *An Essay on the Principle of Population* (Cambridge University Press, 1992), 89.

3. Michael Hudson, *Killing the Host: How Financial Parasites and Debt Destroy the Global Economy* (ISLET-Verlag, 2015), 383.

4. Andy Stern with Lee Kravitz, *Raising the Floor: How a Universal Basic Income Can Renew Our Economy and Rebuild the Amrican Dream* (New York: PublicAffairs, 2016), 39.

5. Ibid., 64.

6. The closest thing America has to a basic income is the Earned Income Tax Credit (EITC), introduced in 1975. To qualify, people must have earned income that falls below a certain level and also have either a qualifying child or be 25 to 65 years old. It is a miserly program: in 2019 the maximum credit for someone with two qualifying children was $5,828.

7. *Walden,* "Conclusion" chapter.

8. Donald Winch, ed., *An Essay on the Principle of Population* (Cambridge University Press, 1992), 198.

9. Edward Abbey, *Confessions of a Barbarian: Selections from the Journals of Edward Abbey*, ed. David Petersen (Boulder, CO: Johnson Books, 1994), 304.

10. During the economic panic of 1893, Detroit's mayor, Hazen Pingree, allocated 945 plots on vacant land to Detroit's destitute. These plots were called Pingree's potato patches.

11. Thomas Jefferson, *Notes on the State of Virginia* (1787), query XIX. Think of Jefferson's slaves when you read that last sentence.

12. Wendell Berry, *What Matters: Economics for a Renewed Commonwealth* (Berkeley, CA: Counterpoint, 2010), 64–65.

13. Wikipedia article entitled "Social Security (United States)" at http://en.wikipedia.org/wiki/Social_Security_%28United_States%29.

14. Michael Hudson, *Killing the Host: How Financial Parasites and Debt Destroy the Global Economy* (ISLET-Verlag, 2015), 375.

15. *Capital in the 21st Century*, p. 275.

16. Presidential inauguration speech, January 20, 1961.

Reducing Crime

Implementing a universal basic income in conjunction with a credit-based population reduction program would be the most effective way to reduce crime. Such a program would reduce crime in several ways. First, the existence of a large (eventually universal) DNA database would result in the swift identification and apprehension of many more criminals. This would not only prevent them from committing more crimes but would have a strong deterrent effect on those contemplating crime. It would also greatly reduce the number of convictions of the innocent, thereby increasing convictions of the guilty. And crimes committed by the poor would be further reduced by the universal basic income and the prevention of excessive childbearing (thanks to the two-credit system). The poor would also have the option to make a great deal of money by selling one or both of their population credits. Eliminating illegal immigration would further reduce crime by opening up many more job opportunities for the poor and teenagers and for those with criminal records (who must now contend with severe employment discrimination). By reducing the teen birth rate to near zero, child abuse and the associated need to place children in foster care would diminish. As neglected children become rare, the social costs of neglect, including crime, would diminish. The freedom to abort an unwanted fetus would also help to reduce crime, because maternal rejection is an important factor in antisocial behavior and violence. Easily accessible abortion would also nearly eliminate infanticide.[1]

A population-stabilization plan would further reduce crime by reducing the number of impulsive people in the population. Impulsive people would likely sell their credits quickly in order to enjoy spending the money. This would deprive them of the opportunity to have children, thereby preventing them from passing on their genes for impulsiveness. The result would be a rapid reduction in the number of impulse-related crimes such as shoplifting, domestic abuse, rape, and murder.

Scientists are discovering that there are many areas of the brain involved in antisocial violence:

> We've also seen that there is not one but multiple brain areas which, when dysfunctional, can predispose one to violence. It's not just the dorsal and ventral regions of the prefrontal cortex that are dysfunctional, but also the amygdala, the hippocampus, the angular gyrus, and the temporal cortex. Yet future research will show it's even more complicated. The antisocial brain is a patchwork of dysfunctional neural systems and we are only just on the threshold of putting together these pieces to better understand it.
>
> —Adrian Raine[2]

These defects in brain anatomy are, to a large extent, genetic:

Taken together as a whole, these studies converge on a simple truth that even the strongest critics of genetic influences in violence are finding harder to resist—genes give us half the answer to the question of why some of us are criminal, and others are not.[3]

A meta-analysis of 103 studies found that 65 percent of aggressive antisocial behavior is heritable.[4] However, genetic predispositions can be strongly modified by environment. Positive influences such as a loving family and adequate food and resources can minimize and even override antisocial genes.[5]

One factor that contributes to the creation of criminals is the absence in infancy of a continuous loving bond with the mother.

What transpires is that there is a critical period early in life when being connected with the mother really counts. In humans this starts at about six months and ends after about two years. For this reason breakage of the mother-infant bonding process for at least four months in the first year of life—as experienced by some of our Copenhagen babies—freezes the social-interpersonal development of the infant. That freezing results in the glacial, emotionless psychopath that we see in adulthood.[6]

—Adrian Raine[6]

Another factor known to predispose children to hyperactivity and aggression is a mother who smokes during pregnancy.[7] The culprit is nicotine, which crosses the placental barrier, resulting in higher testosterone levels in fetuses and smaller brain mass in critical areas that regulate aggression.[8] The severity of the effect is dose dependent: the greater the intake of nicotine during pregnancy, the greater the likelihood the child will become hyperactive and aggressive.[9] In the United States, about one-quarter of pregnant women smoke, so this is a major problem.

Pregnant women also harm their fetuses when they consume alcohol. In extreme cases this results in fetal alcohol syndrome, whose symptoms include a shrunken brain and distinctive facial features. As with smoking, fetal alcohol damage is dose dependent: the more a pregnant woman drinks, the more the fetus is damaged. One way to reduce the incidence of alcohol-damaged children would be to mandate long-term contraception for every woman who has already produced an alcohol-damaged child. She should not be allowed to become pregnant again until she overcomes her addiction. One simple step that would help to reduce the damage caused by alcohol and tobacco would be to completely ban their advertising.

Exposure to lead may also play a significant role in crime. Lead is neurotoxic and has deleterious effects on brain development. It results in smaller brain volume, especially in the frontal cortex, and this lowers intelligence and increases aggression. Even hamsters become more aggressive when experimentally exposed to lead.[10] The lower intelligence produced by lead exposure leads to social and economic discrimination, further increasing the likelihood of aggression and crime. Lead poisoning primarily impacts children who live in poverty. They consume lead by eating peeling paint in old houses and by breathing dust contaminated with lead. Leaded gasoline was formerly a major contributor of lead in cities. The large drop in violent crime in the United States since 1993 may be partially due to the elimination of leaded gasoline 20 years earlier. In fact, one study found that the elimination of leaded gasoline could account for 91 percent of the variance in violent offending.[11] Lead can also leach from water pipes if the water is too acidic, as happened in Flint, Michigan, in 2014, when the city was switched by a state-appointed administrator to a new water source without properly treating the water to reduce corrosion. Although government agencies have set "safe levels" for lead exposure, there is no safe level.

Poor nutrition is another factor in violent crime. Fetuses, children, and adults need adequate amounts of zinc, iron, riboflavin, fiber, and omega-3 fatty acids (DHA and EPA).[12] Omega-3 fatty acids are especially important: multiple studies have shown that they can substantially reduce levels of violence.[13] Consuming too much sugar also plays a role in aggression, especially if consumed in the absence of fiber-rich foods. Fiber slows the absorption of sugar, preventing it from rushing into the bloodstream and triggering an excessive surge in insulin. Too much insulin lowers blood sugar excessively, causing people to become nervous and irritable, which can lead to angry outbursts. A study in Finland found that low blood sugar can reliably forecast which criminals go on to commit more crimes.[14] A study in Wales found that children who eat sweets every day were three times more likely to become violent by age 34.[15]

Some children have chronically slow brain waves (delta and theta waves), which indicate an underaroused brain. These children have difficulty focusing on a task and continually seek excitement in order to achieve a normal state of arousal. Unfortunately, they may find their excitement in crime. Case studies have demonstrated that a prolonged course of biofeedback can reset the brains of these children, causing the cortex to resume normal alpha and beta waves. This frees the child from the need for continual stimulation, allowing him or her to focus on the task at hand and removing the attraction of crime.[16]

There is evidence that "mindfulness" meditation may beneficially alter brain structure. It may have great potential to reduce violence. But so far no randomized controlled trials have been conducted to determine its effectiveness.[17]

Most crime committed by the mentally ill could easily be prevented if their calls for help (and the calls for help by their relatives and friends) did not go unheeded. In the United States, mental health services are poorly funded, due to miserly state legislatures. There are far too few mental hospitals.

Better protection of children from abusive parents would go a long way toward preventing crime. Many children are beaten on their heads and thrown around violently, or they may be violently shaken as babies. This causes permanent brain damage, which leads to learning disabilities and psychopathy. One reason parents shake their babies is that the babies may be crying incessantly from the pain of colic. Colic is often caused by the proteins in cow milk, which the baby can acquire even from breast milk if the mother consumes dairy products.[18] Breastfeeding mothers should instead consume milks made from seeds and nuts.

A proven way to reduce the likelihood that impoverished children will become criminals is to assign a nurse to a low-income pregnant woman. The nurse makes periodic home visits during pregnancy and for two years following birth. She teaches the mother the importance of avoiding tobacco and alcohol and teaches her how to treat and care for a baby, placing particular emphasis on love, close physical contact, proper nutrition, and nonviolent discipline. Children in one study were observed over the course of 15 years. Those who received the nurse visits had a 53 percent reduction in arrests and a 63 percent reduction in convictions. Not only did the program reduce suffering, but it more than paid for itself.[19]

Experiments have shown that raising dopamine levels and reducing serotonin levels increases aggression. Serotonin helps people control their tempers: when serotonin levels drop, tempers flare. The serotonin transporter gene has two variants that lower serotonin levels. One variant (the short-allele variant) promotes impulsive, hotheaded aggression, while the long-allele variant promotes cold-blooded, carefully planned psychopathic behavior.[20] Probably at least 10 percent of aggression is due to low serotonin levels.[21]

A poorly functioning prefrontal cortex is another important factor in antisocial and violent behavior.[22] Many studies support this conclusion. PET scans of murderers show that they have low glucose metabolism in the prefrontal cortex compared to controls. This results in less self-control, hence more angry outbursts, more violence, more risk-taking, more rule breaking, and more irresponsibility. A poorly functioning prefrontal cortex also contributes to poor social skills and lower intelligence, which in turn lead to further social marginalization.

People with antisocial personality disorder tend to have resting heart rates that are considerably lower than those of normal people.[23] This holds true for both sexes and all age groups. Studies of twins have demonstrated that heart rate is inherited. A low resting heart rate correlates with fearlessness, lack of empathy, and a chronically low state of arousal, which leads to a quest for stimulation—too often through crime and violence.

In unstressed research subjects, resting heart rate accounts for 5 percent of the difference among people in antisocial behavior, while in stressed subjects it accounts for 12 percent. Males generally have slightly lower resting heart rates than females, which accords with the greater antisocial behavior of males. Resting heart rate is at its lowest during adolescence, when violence and thrill seeking are at their peak. A low heart rate in childhood is an even better indicator of future antisocial behavior than having a criminal parent. Researchers have found that children who are strong stimulation seekers at age three are more likely to be aggressive at age 11. Just as a low resting heart rate is correlated with antisocial behavior, a high resting heart rate is correlated with prosocial behavior. Stimulants that increase heart rate decrease the incidence of antisocial behavior. This raises the possibility of using stimulants to reduce crime rates and shrink prison populations. It also suggests that psychotherapy may be of little value for people who have low resting heart rates.

Psychopaths have no conscience and feel no empathy, shame, or remorse. They move frequently and have no long-term goals. Despite having no interest in long-term relationships, they are socially adept. PET scans consistently show that psychopaths have a poorly developed orbitofrontal cortex, one of the seats of conscience. And their amygdala volume is 18 percent smaller than average.[24] The amygdala is involved in fear conditioning. Without fear conditioning, the conscience cannot develop properly, and impulses are harder to control. The poor development of the amygdala in psychopaths may be genetic or due to damage during fetal development. Absence of maternal love during childhood can also interfere with proper development of the amygdala.

The hippocampus (a small brain structure lying behind the amygdala) is also involved in violence. In psychopaths, volume reductions have been observed in areas of the hippocampus involved in fear conditioning. However, the right hippocampus in psychopaths is much larger than the left.[25] Moving rats or humans from one home to another during childhood tends to enlarge the right hippocampus. This suggests that parents should remain in one home while their children grow up. Fetal alcohol exposure also greatly enlarges the right hippocampus.

The corpus callosum, which connects the two hemispheres of the brain, is much larger in psychopaths and also longer and thinner. This likely contributes to their above-average conversational skills. The striatum of psychopaths is also about 10 percent larger than normal, and this may be a factor in their greater reward-seeking behavior.[26] Normal people are able to control their desires for money, sex, and power but psychopaths seem unable to. They are driven to achieve their goals and have no conscience to get in the way. Even genetically normal people can be transformed into psychopaths by damage to the ventral prefrontal cortex due to blows or tumors.[27]

The US has a far higher rate of gun violence than any other country. Between 1968 and 2011, 1.4 million Americans were killed by guns. More than a third of these deaths were homicides, while the rest were suicides or accidents. There have been many failed proposals to address this problem, but there is a simple and practical solution that is never mentioned. That would be a federal law requiring people who wish to purchase a gun to present affidavits from 10 citizens of good standing who personally know the purchaser attesting that he or she is a person of good moral character who would use the weapon legally and care for it responsibly. Once the affidavits are approved by a court, the court would forward the approval to a gun manufacturer. The manufacturer would then make the weapon. This process would eliminate gun dealerships, since customers would deal directly with the manufacturer through the court (much as Tesla by-passes dealerships by dealing directly with its customers). We should also ban all advertising of guns. As for the people who already own guns, they should be required to retroactively present 10 affidavits within a prescribed period. Generous rewards would be offered for tips about people who have illegal guns, and those guns would be confiscated and destroyed. Gun ownership, like possessing a driver's license, should be a privilege, not a right. Contrary to the dyslexic opinion of the US Supreme Court, the US Constitution does not grant citizens the right to bear arms. The Second Amendment grants the right to bear arms to well-regulated state militias. It says noth-ing about the right to bear arms by private citizens. Therefore, in accordance with the Tenth Amendment (which grants the powers not delegated to the federal government to the states), it is up to each state to determine how (or whether) to regulate private gun ownership. For example, a state could legally require all its citizens to own firearms. Alternatively, it could ban all private gun ownership. A state could even establish reasonable firearm regulations that the public would support.

Sometimes criminals aren't even criminals but innocent people who were wrongfully charged and convicted. In the United States, DNA testing has revealed that convictions of the innocent are not rare. To convict someone who is innocent is both a crime against that person and a crime against the original victim and society (because the real culprit remains free to commit more crimes). Many con-victions of the innocent are due to sloppy investigations, erroneous witness identifications, the suppression of exculpatory evidence by prosecutors, coerced confessions, bogus forensic "science," and the false testimony given in return for a promise of leniency. Less commonly, wrongful convictions stem from false accusations made by someone seeking personal revenge.

Witness misidentifications could be greatly reduced by completely eliminating the traditional police lineup, and replacing it with this procedure:

- The subjects should be presented one at a time to the witness, rather than simultaneously in a lineup.
- All the subjects should roughly match the witness's description (and if one of them will be handcuffed, all of them should be handcuffed).
- The policeman conducting the examination should not know who the suspected criminal is, lest he or she intentionally or unintentionally provide subtle hints to the witness.
- Witnesses should rank each subject by how closely he or she resembles the actual criminal as best they can recall.
- The whole identification process should be videotaped.

At least some of these lineup reforms have already been implemented in 11 states and some cities. We should always be skeptical of witness identifications and rely whenever possible on solid forensic evidence. Police forces should receive the funding to develop better forensic evidence.

The withholding of exculpatory evidence by prosecutors should be treated as an extremely serious crime. Prosecutors need to understand that their responsibility to the public is to secure justice, not to rack up convictions like points in a game.

One of the most effective ways to reduce a broad spectrum of crimes would be to reduce our own materialism. The materialistic values of criminals reflect the materialistic values of their society. Criminals want to enjoy the same respect that society confers upon everyone else who has money and coveted possessions.

> With gold and jade in the hall, the house isn't safer.
> —*Tao Te Ching*[28]

Another way we could immediately and drastically shrink our court dockets and prison populations would be to abolish victimless crime laws. Thomas Jefferson denounced such laws, declaring that "the legitimate powers of government extend to such acts only as are injurious to others."[29] Jefferson's sentiment was echoed by Supreme Court Justice James Wilson (a signer of the Declaration of Independence and a great legal scholar) who said "Every crime includes an injury: every injury includes a violation of right."[30] John Stuart Mill elaborated: "…the two essential ingredients in the sentiment of justice are, the desire to punish a person who has done harm, and the knowledge or belief that there is some definite individual or individuals to whom harm has been done."[31]

Although victimless crimes injure no one, victimless crime *laws* have injured many. These laws have damaged the lives of millions of citizens through senseless arrests and prosecutions, which have caused lasting harm to the victims and have cost taxpayers an immense amount of money for policing, courts, prisons, and probation. The United States imprisons a far larger percentage of its population than any other nation: about 716 per 100,000, compared to 118 in Canada and 210 in Mexico.[32] Many of these US prisoners are nonviolent drug offenders. In 2014, the US had 105,284 federal prisoners and 216,153 state prisoners doing time for drug offenses.[33] In addition, there are millions more who have already served their sentences but must now endure a lifetime of job and housing discrimination and suffer denial of various government benefits. In some states, they are also stripped of their right to vote (in patent violation of the 14th and 15th Amendments of the US Constitution). Their children and other family members also suffer deep harm, as does the entire society from the contempt shown for personal liberty.

Another way victimless crime laws cause injury and death is by enabling the existence of drug cartels. In the four years from 2006 to 2010, some 30,000 Mexicans were killed in drug-trade violence, and thousands more have died since. The ultimate responsibility for this carnage lies with the US Congress and US state legislatures, which have enabled it through their "victimless" crime laws. The governments of Latin America also deserve blame. If they had common sense and a modicum of courage, they would legalize drugs. (In 2021, Mexico led the way by legalizing recreational use of cannabis.) Not only would this end drug violence, but it would attract more tourists.[34]

Three of America's recent presidents—Obama, Bush Jr., and Clinton—all violated the nation's drug laws as young men.[35] Yet when they became president, they did nothing to reform these laws. At the very least, they should have removed marijuana and psychedelics from Schedule I (one of the five official drug classes). This would have corrected a legal travesty because none of these drugs meets the three criteria that the government established for Schedule I; to wit: (1) no accepted medical use in the US, (2) high potential for abuse (i.e., addictive), and (3) unsafe to use even under medical supervision. These presidents should also have granted pardons to all the people convicted in the federal court system of nonviolent drug "offenses" (in other words, all the people who happened to be unluckier than these three hypocritical presidents).

Not many people realize that legislative reform of drug laws could be brought about in a matter of weeks if those arrested for drug offenses would demand their constitutional right to a jury trial and able legal representation. This would immediately gridlock the legal system and force governments to reform their drug laws. Consider this: in New York City in 2012, there were 50,000 drug arrests! If all these cases went to trial, it would require the recruitment of hundreds of thousands of jurors (many of whom use drugs themselves) and an army of court-appointed attorneys. But these cases rarely go to trial because defendants are intimidated into plea bargaining. Prosecutors achieve this intimidation by charging misdemeanors as felonies and by charging defendants with as many different offenses as prosecutors can squeeze out of the law. Another way prosecutors intimidate defendants into plea bargaining is by denying them bail. This forces defendants to remain in overcrowded jails (sometimes in solitary confinement) until they finally break down and consent to a plea deal. A 16-year-old boy in New York City named Kalief Browder, who refused to accept a plea deal after he was accused of stealing a backpack, was held in solitary confinement for two years without trial. Kalief later killed himself. No one in the New York judicial system was ever held accountable for his imprisonment, despite the fact that failing to give him a speedy trial was a gross violation of the Fifth Amendment of the US Constitution.[36] Defendants like Kalief, who dare to assert their legal right to have a public defender, quickly discover that public defenders are overworked and often incompetent. They cannot provide a capable defense. This is another violation of the Fifth Amendment. Moreover, 44 of the 50 states plus the District of Columbia have adopted the despicable practice of billing impoverished defendants for the costs of their public defender. And when defendants cannot pay, they may be thrown in jail (debtors' prison).[37] This is yet another way that America's powerful oppress the powerless.

By means of intimidation, prosecutors in the US are now able to get 95 percent of defendants to plea bargain. Lawyers have colluded in this corrupt system because it is easier and more lucrative for them to arrange plea bargains than to do the hard work of preparing a case and presenting it to a jury. The Founding Fathers were right to insist that jury trials are the best defense against tyranny. The present state of our legal system would have appalled them.

The most effective long-term solution to victimless crime laws would be to outlaw them with a constitutional amendment. The language of the amendment should be modeled after these words of John Stuart Mill:

> The sole end for which mankind are warranted, individually or collectively, in interfering with the liberty of action of any of their number, is self-protection. That the only purpose for which power can be rightfully exercised over any member of a civilized community, against his will, is to prevent harm to others. His own good, either physical or moral, is not sufficient warrant. He cannot rightfully be compelled to do or forbear because it will be better for him to do so, because it will make him happier, because, in the opin-

ion of others, to do so would be wise, or even right...
The only part of the conduct of anyone, for which he
is amenable to society, is that which concerns others.
In the part which merely concerns him, his independence is, of right, absolute. Over himself, over his own
body and mind, the individual is sovereign.[38]

Such an amendment would have been included in the original
US Bill of Rights if not for the fact that so many of the Founding
Fathers were slave owners. Their slaves were under a life sentence of
involuntary servitude, even though they had never harmed anyone.

The legalization of drugs does not mean that they would be sold
freely from street corner kiosks. Sales should be restricted to adults
who have passed an exam to demonstrate that they understand the
physiological, psychological, and social risks—and benefits—that
each drug presents. Female users of addictive drugs (including tobacco and alcohol) should always use long-term, fail-safe contraception
(such as IUDs). In Portugal, all drugs are now legal, and the Portu-
guese system is working admirably. Addiction rates and HIV rates in
Portugal have gone way down.

Finally, the Black Lives Matter movement has underscored the
need for police reform. One of the best ways to reduce potentially
dangerous confrontations between police and citizens would be to
ban traffic stops. Traffic stops could be eliminated by establishing a
nationwide network of computerized traffic cameras, which would
identify violating vehicles and automatically issue tickets or court
summonses to the owner of the vehicle. This would remove police
from traffic enforcement and would protect the police themselves
(and innocent bystanders) from high-speed chases. It would also save
municipalities a large amount of money, since they could substan-

tially reduce the size of both their police force and its vehicle fleet. Money would also be saved because police would not have to spend time testifying in traffic court.

Another helpful police reform would be to better screen applicants before hiring them. There should be a national police database, so that if an officer is fired from one municipality, other municipalities could learn about it and avoid hiring the troublemaker. Police applicants should also be required to take psychological tests to identify those with strong racial bias or an inclination to bully and abuse. There are two other reforms that would greatly help to end police abuse: first, police officers should always live in the neighborhoods they police (ideally, this would be the same neighborhood where they grew up); second, the racial and cultural composition of the police force should reflect the population being policed.

Notes

1. Adrian Raine, *The Anatomy of Violence: The Biological Roots of Crime* (New York: Vintage Books, 2013), 22.

2. Ibid., 99.

3. Ibid., 47.

4. Ibid., 42.

5. Ibid., 246–47.

6. Ibid., 191.

7. Ibid., 198.

8. Ibid., 197–198, 200–201.

9. Ibid., 199. Smoking mothers have higher testosterone levels, which results in higher testosterone levels in the developing brains of their

fetuses. This creates more highly masculinized brains, which would explain why the children of smokers are more aggressive. See p. 197 of Raine's book for more about this.

10. Ibid., 225.

11. Ibid., 226.

12. Ibid., 206–241.

13. Ibid., 296.

14. Ibid., 222.

15. Ibid., 223.

16. Ibid., 274.

17. Ibid., 297–300.

18. See https://nutritionfacts.org/video/treating-infant-colic-by-changing-moms-diet/.

19. Adrian Raine, *The Anatomy of Violence,* 277–78.

20. Ibid., 58.

21. Ibid., 57.

22. Ibid., 67–68.

23. Ibid., 108–110.

24. Ibid., 161.

25. Ibid., 163.

26. Ibid., 167.

27. Ibid., chapter 5: "Broken Brains."

28. Lao Tzu, *Tao Te Ching: A New Translation*, Sam Hamill trans. (Boston: Shambhala Publications Inc.), chapter 9.

29. Thomas Jefferson, *Notes on the State of Virginia* (1787), query on religious freedom.

30. James Wilson, *Lectures on Law* (1790–92), III, ii.

31. John Stuart Mill, *Utilitarianism* (1863), chapter 5.

32. Roy Walmsley, "World Prison Population List," 10th edition. United Kingdom Home Office Research, 2013. See www.prisonstudies.org/sites/default/ files/resources/downloads/wppl_10.pdf.

33. See www.drugwarfacts.org/cms/Prisons_and_Drugs#sthash.GhwKY8dZ.dpbs. Also see http://www.bjs.gov/index.cfm?ty=pbdetail&iid=5387.

34. Legalizing drugs in the US would cause Mexico to lose one of its largest sources of hard currency, which would increase emigration from Mexico. So the US should work closely with Mexico to help it withdraw from its drug-money dependency without increasing emigration.

35. Clinton, Bush Jr., and Obama have all admitted to using marijuana (though Clinton laughably claimed he didn't inhale). Bush made his admission in an interview with Doug Wead (see http://www.nytimes.com/2005/02/20/politics/20talk.html).

36. See http://www.newyorker.com/magazine/2014/10/06/before-the-law. Alexis de Toqueville, in *Democracy in America*, observed that "The poor man cannot always find the money for bail, even in a civil matter. If he is obliged to await justice from prison, his enforced constraint soon reduces him to a wretched state. By contrast, the wealthy man always evades prison in civil matters."

37. See http://www.npr.org/2014/05/19/312158516/increasing-court-fees-punish-the-poor.

38. John Stuart Mill, "The Contest in America," *Harper's Magazine*, April 1862, vol. 24, issue 143, 677–84.

Sensible Alternatives to Prison

Prisons are a major drain on taxpayers, and they waste human potential. They also largely fail to prevent further crimes once prisoners are released. In the state of Michigan, 40 percent of released prisoners are returned to prison within three years.[1] There are practical ways to reduce prison populations without endangering public safety. Here are a few of them:

1. Take practical steps to prevent crime, such as eliminating poverty, developing mentoring programs for at-risk youth, eliminating lead poisoning, and reducing birth complications (which can starve infant brain cells of oxygen, resulting in permanent brain damage that is strongly correlated with violent crime later in life).[2]
2. Cease criminalizing behavior that has no victims (and make addiction treatment readily available).
3. Use alternative sentences that would allow the convict to make restitution for the crime (provided that would pose no serious risk to public safety).
4. Give employers an economic incentive to hire nonviolent convicts as an alternative to imprisonment, and require that convicts use part of their salary for restitution.

5. Adequately fund mental health services so that prisons will no longer be used to warehouse America's mentally ill (who are legion).

To provide an incentive to employers to hire convicts, the state should offer the employer the money it would otherwise spend to incarcerate the convict. Currently, US states pay an average of $79 per day to keep each prisoner. This amounts to $28,835 per year.[3] Although paying employers this money would not reduce the state's expenditures, it would increase the state's income, because employed convicts would pay state taxes. The employer would benefit from the money received from the state, and from the profit produced by the convict's labor. Convicts, too, would get a double benefit: they would have their freedom and have a steady job. Employed convicts who wish to move to a different job in the same state would be free to do so, taking their state subsidy along with them. The employment of convicts would not cause resentment among other workers if the population were shrinking, because a shrinking population increases demand for labor.

In the case of prisoners who are too mentally impaired and dangerous to be released, the establishment of prison agricultural farms would be a sensible alternative.

To better integrate convicts into society, there should be a jubilee law to eliminate criminal records and restore full civil rights to any convict who stays out of trouble for, say, two years after being released from prison. The prisoners who founded modern Australia demonstrated that if convicts are given an opportunity to make a legitimate living, they are happy to do so. Charles Darwin observed this firsthand:

...as a means of making men outwardly honest,—of converting vagabonds, most useless in one hemisphere, into active citizens of another; and thus giving birth to a new and splendid country—a grand centre of civilization—it has succeeded to a degree perhaps unparalleled in history.[4]

In the case of male child molesters and rapists, the only practical solution is to drastically lower their testosterone levels to prepubertal levels with mandatory hormone treatments or castration. There are several chemicals that can accomplish this, including medroxyprogesterone (Depo-Provera) and cyproterone acetate. Chemical treatments should be supplemented with psychotherapy in cases where it might have benefit. One caveat about chemical castration is that it is only effective against sex crimes motivated by lust, not those motivated by hatred.

Castration would benefit everyone: the victim would know that his or her rapist could never rape again and that his castration would serve as a deterrent for would-be rapists; society would benefit in two ways: first, by not having to spend a large amount of money to house the rapist in prison only to see him commit more rapes once released; second, by preventing the rapist from transmitting his genes to another generation—genes that may have predisposed him to rape. The rapist would also benefit by being able to get on with his life and enjoy an additional 14 years of life span. (Castration is, by far, the most effective way to extend male life span.) The rapist would, however, have to pay the expenses incurred by the victim for psychological counseling and legal expenses, as well as the time she (or he) lost to courtroom proceedings.

In Germany, rapists have the option of surgical castration. Those who have chosen castration have a recidivism rate for sexual offenses of about 3 percent compared to 46 percent for the noncastrated.[5] There is some evidence that surgical castration is even more effective than chemical castration.[6] However, a small minority of men who have undergone castration still retain some sexual desire and can still get erections.[7] Therefore, prior to release from detention, castrates should be tested to make sure they can no longer function sexually. In addition, they should remain indefinitely under parole supervision so they can be regularly tested to ensure they are not using exogenous testosterone to boost their sex drive. Castration sometimes has undesirable side effects such as weight gain, breast development, thinning of body hair, and depression, all of which provide a motive for castrates to obtain exogenous testosterone.

Using the measures described above to counteract violence would replace our present vindictive approach to crime with a pragmatic one. It would improve public safety by reducing recidivism and it would lower taxes by reducing government expenditures. If the public understood that a pragmatic crime policy is far more effective and far less expensive than a vindictive one, most people would support it. Unfortunately, there are now many powerful economic interests that feed on America's policy of mass incarceration. These include prison guards, prison bureaucrats, prison architects, prison builders, prison food suppliers, lawyers, judges, court officials, probation officers, manufacturers of kits for drug testing, drug lab technicians, and the thousands of Americans who earn their livelihood by watching their fellow citizens urinate into test vials—all of these people would suffer losses from legal reform. So be it.

Notes

1. *LSA Magazine*, Spring 2011 (College of Literature, Science and the Arts, University of Michigan), 23.

2. Adrian Raine, *The Anatomy of Violence: The Biological Roots of Crime* (New York: Vintage Books, 2013), 188–89.

3. *LSA Magazine*, Spring 2011 (College of Literature, Science and the Arts, University of Michigan), 25.

4. Charles Darwin, *Voyage of the Beagle*, chapter 19.

5. Adrian Raine, *The Anatomy of Violence,* 284.

6. Ibid., 286.

7. Linda E. Weinberger et al. "The Impact of Surgical Castration on Sexual Recidivism Risk Among Sexually Violent Predatory Offenders." *J Am Psychiatry Law* 33 (2005): 16–36.

Part 6:

Health Care Reforms

Preventing Disease and Providing Health Care

> Health is a precious thing, and the only one, in truth, which deserves that we employ in its pursuit not only time, sweat, trouble, and worldly goods, but even life; inasmuch as without it life comes to be painful and oppressive to us.
>
> —Michel de Montaigne[1]

Medicare and Medicaid currently consume more money than the US military budget. Preventing chronic diseases would free up a vast amount of money that could easily cover the cost of government purchases of population credits and a universal health care system. It would also eliminate an immense amount of suffering and strengthen the economy.

Under the two-credit system, reductions in the mortality rate brought about by improvements in disease prevention and treatment would *not* lead to an increase in population size. That is because additional credits would be retired to compensate for the reduced mortality. This credit retirement would be handled automatically by computers linked to a continually updated vital statistics database.

The leading causes of premature mortality in the US (ignoring COVID-19 whose impact will likely prove temporary) are these:[2]

1. Coronary heart disease (heart attacks)—375,000 deaths annually
2. Lung diseases (lung cancer, chronic obstructive pulmonary disease, asthma)—296,000 deaths annually
3. Improper medical care (wrong prescriptions, unnecessary or incompetent surgeries, bad drug reactions, hospital-acquired infections)—225,000 deaths annually
4. Brain diseases (stroke and Alzheimer's)—214,000 deaths annually
5. Digestive cancers (colorectal, pancreatic, esophageal)—106,000 deaths annually
6. Infections (respiratory and blood)—95,000 deaths annually
7. Diabetes—76,000 deaths annually (the actual number is much higher because when diabetes is an underlying cause of death it is rarely listed on death certificates)
8. High blood pressure—65,000 deaths annually
9. Liver disease (cirrhosis and cancer)—60,000 deaths annually
10. Blood diseases (leukemia, lymphoma, myeloma—56,000 deaths annually.
11. Kidney disease—47,000 deaths annually
12. Breast cancer—41,000 deaths annually
13. Suicide—41,000 deaths annually (in 2018, 24,432 were by firearms)
14. Prostate cancer—28,000 deaths annually
15. Parkinson's disease—25,000 deaths annually
16. Homicide—21,570 deaths in 2020 (nearly all by firearms)

By adopting health-promoting diets, ending tobacco use, moderating alcohol consumption, reforming medical procedures, properly vetting gun purchasers, and establishing nationwide public health policies, we could eliminate most of these premature deaths.

Although most people are well aware that tobacco and alcohol abuse are harmful, there is less appreciation of the dangers of deleterious foods. The medical profession is partly to blame for this ignorance because it has focused entirely on treatment rather than prevention. This is understandable, since treatment, under our present medical system, is lucrative while prevention is not.

Another source of dietary ignorance is books that promote fad diets. One author of such books, Dr. Robert Atkins, made hundreds of millions of dollars by assuring people that they could eat their favorite fatty foods with impunity.[3] Atkins himself was obese and died after collapsing and striking his head. He had a medical history of hypertension and congestive heart failure and had previously suffered a heart attack.

Unfortunately, the nutritional guidelines developed by the US Department of Agriculture are biased due to a conflict of interest. On the one hand, the Department of Agriculture is responsible for establishing sound nutritional guidelines, but on the other hand, it is required to promote the interests of the meat, poultry, and dairy industries. In order to eliminate this conflict, responsibility for nutritional guidelines should be shifted from the Department of Agriculture to unbiased nutritional scientists.[4] *Unbiased* means scientists who do not moonlight as consultants for the meat, dairy, egg, fish, and sugar industries.

To see how biased US nutritional guidelines are in favor of meat and dairy products, we need only compare them to the dietary guidelines of India. Indian guidelines recommend protein from plants rather than meat or dairy products. As a result, the carbon footprint of the recommended Indian diet is far smaller than that of the recommended American diet.

The public's ignorance and confusion about healthful diet is tragic, because nutritional scientists and informed physicians have known for more than two decades that if people adopted a diet rich in whole grains, legumes, vegetables, fruits, and nuts they would see the following benefits:

- the virtual elimination of heart attacks and strokes (which are the first and fourth leading causes of death in the US);[5]
- a dramatic decrease in the incidence of diabetes;
- a large decrease in cancer (colorectal, prostate, breast,[6] liver, blood, and probably others);
- the gradual and permanent loss of excessive weight (especially if combined with exercise);
- a reduction in osteoarthritis;
- a reduction in osteoporosis (hence fewer broken bones in the elderly);
- a reduction in multiple sclerosis;
- a reduction in Alzheimer's disease;[7]
- and a reduction in vision loss due to macular degeneration.[8]

As Lucretius observed:

It was lack of food that killed men in the past. Today it's overindulgence and excess.[9]

Additional benefits of a plant-based diet are that people would look and feel better, and the economy would benefit from lower medical costs, fewer worker absences due to illness, and fewer breadwinner deaths. And there would be a vast reduction in emissions of CO_2 and methane, which would do even more to reduce climate change than the elimination of internal combustion vehicles and coal-fired power plants. Without all these burdensome costs, the products we manufacture could be sold more competitively both at home and abroad, while simultaneously reducing unemployment and the trade deficit. Better health would increase the wealth of the average American by reducing sick days, reducing the cost of medical insurance, and reducing taxes thanks to plummeting costs for Medicare and Medicaid. Adopting a plant-based diet would also reduce soil erosion and reduce the amount of fertilizer and pesticides applied (resulting in less water pollution and lower concentrations of toxins in our bodies).

A plant-based, low-fat diet not only prevents the buildup of artery-clogging plaque that causes heart attacks and many strokes, but actually reverses this condition in most people. This has been established by solid research accepted by the scientific community. Among the best books on how to use diet to prevent and reverse heart disease are those by Caldwell Esselstyn[10] and Dean Ornish.[11] Both are medical doctors with impeccable credentials. And both have done original human research that has been published in reputable medical journals. In *Prevent and Reverse Heart Disease*, Dr. Esselstyn includes impressive before-and-after angiograms of coronary arteries showing the effectiveness of a plant-based diet in clearing arteries of plaque, even in the absence of regular physical exercise. Esselstyn's website is www.heartattackproof.com.

A plant-based diet can also largely reverse type 2 diabetes without drugs and can reduce insulin requirements for those who have type 1 diabetes if their diet is rich in whole carbohydrates. It may even prevent type 1 diabetes altogether.[12] A useful book for diabetics is *Dr. Neal Barnard's Program for Reversing Diabetes* (2007, Rodale). A plant-based diet can also reduce cancer: in 2015, the World Health Organization classified processed meat (hot dogs, salami, bologna, corned beef, pepperoni, bacon, and sausages) as carcinogens, because they cause colorectal cancer and likely also cause stomach cancer. WHO warned that other forms of "red meat" are probable carcinogens, because their consumption is associated with colorectal, pancreatic, and prostate cancers. This relationship is dose dependent: the more meat consumed, the greater the risk.[13] Plant-based diets help protect against all cancers, but the protective effect is especially pronounced for the blood cancers: leukemia, lymphoma, and myeloma.[14]

When a group of olive baboons in Kenya adopted the modern Western diet, their health seriously degenerated:

> Every lodge and tourist camp thus had a resident troop of baboons living on the garbage. The animals would no longer go out to forage, but would sleep in the trees above the dump, snoozing late and waiting for the daily garbage drop. I had even studied the metabolic changes that the baboons living at another lodge had gone through as they became garbage eaters, feasting on leftover drumsticks, slabs of beef, rotting dollops of last night's custard pudding. Not surprisingly, cholesterol, insulin, and triglyceride lev-

els rose, other aspects of metabolism went to hell in the same way that ours does when we eat the stuff.[15]

We may be about to learn that the health situation in America is even worse than we already know it to be: recently published research has confirmed the existence of *transgenerational epigenetic inheritance*.[16] This is the transmission via sperm and eggs of genes that have been switched on or off in response to exposure to certain chemicals. Researchers at the University of New South Wales fed slim male rats a high-fat diet typical of what most Americans eat. The rats predictably became fat and developed insulin resistance and glucose intolerance, which are the symptoms of type 2 (adult onset) diabetes. When mated to healthy, slim females, these males sired daughters with insulin resistance and glucose intolerance, *even though the daughters had never eaten a high-fat diet themselves*. If these results from rats apply to humans, Americans are already transmitting their poor health via transgenerational epigenetic inheritance to their children, grandchildren, and great-grandchildren. If that is true, we will eventually have to deal with the ethics and legality of imposing sickness on future generations. Just as we don't allow people to freely transmit the HIV virus, we may have to insist that people change their dietary habits or abstain from childbearing, so they don't transmit diseases like diabetes to their descendants.

Besides transforming the American diet, we need to greatly reduce the unnecessary use of antibiotics. Not only does overuse of antibiotics promote the evolution of antibiotic-resistant bacteria (with potentially lethal consequences), but antibiotics themselves can cause serious long-term health problems. They kill helpful as well as harmful bacteria in our intestinal microbiome (the 12,000 species of bacteria that live in our gut). These beneficial bacteria coevolved with us over millions of years, and they strive to keep us healthy out of self-interest. One way our microbiome keeps us healthy is by helping to prevent obesity. When we batter our microbiome with antibiotics, we lose resistance to obesity. For more than 50 years, farmers have fed antibiotics to their livestock for the specific purpose of fattening them more quickly. The exposure of our children to antibiotics is likely one factor in the obesity epidemic (though bad diet is certainly the main one).[17] Another consequence of overuse of antibiotics in the United States is the elimination of the bacterium *Helicobacter pylori* from the digestive systems of many Americans. Although this has had the beneficial effect of reducing the incidence of stomach ulcers (which are associated with this bacterium), this bacterium also protects against gastroesophageal reflux disease (GERD) and esophageal cancer, which have now reached epidemic proportions in the United States.[18]

Along with improving diet and cutting back on antibiotic use, we could greatly improve our health and greatly lower health care costs by eliminating unnecessary and harmful medical procedures. In a 2010 article in *Newsweek* magazine, Sharon Begley drew attention to the fact that "an estimated one fifth to one third of U.S. health care costs, at least $500 billion a year, goes toward tests and treatments that do not benefit patients."[19] Her article points out that these procedures cause an extraordinary amount of injury and death. "Unnecessary care kills 30,000 Americans every year, estimates Dr. Elliott Fisher of Dartmouth Medical School—and that figure includes only Medicare patients." A similar assessment of the remarkable lethality of American health care appeared in an article of the *Journal of the American Medical Association.*[20] The author, Barbara Starfield, estimated that physician errors, medication errors, and adverse effects from surgery and drugs kill about 225,400 Americans every year. Medical care is now the third leading cause of premature death in America, after heart attacks and cancer and ahead of strokes. Even when useless medical procedures do not result in injury and death, they are still a great drag on the economy, reducing the nation's wealth. Decades ago, tonsillectomies were routinely performed on US children, even though there was no evidence of medical benefit and despite the fact that children sometimes died from surgical complications. But tonsillectomies were profitable for surgeons and hospitals. Another common form of surgery that is often unnecessary is Cesarean sections. In the state of Michigan, in 2008, two-thirds of all babies were delivered surgically rather than born naturally. The main reason for performing these operations is that they are lucrative for obstetricians, anesthesiologists, and hospitals. It is therefore no surprise that they are more common in for-profit hospitals than nonprofit hospitals. In California, for example, for-profit hospitals perform 17 percent more C-sections than nonprofit hospitals.[21] Unnecessary C-sections are not only economically harmful, but they deprive babies of important immunity-building bacteria that they would acquire through natural birth.

Our doctors and hospital administrators try to defend themselves by claiming that their unnecessary medical procedures have some merit or that they are *forced* to perform these lucrative procedures in order to protect themselves from lawsuits by patients. The reason the medical establishment gets away with this mayhem is that they have an extremely powerful lobby: the American Medical Association. This organization and the powerful hospital lobby use payments to legislators to persuade them to do their bidding. That is why the national health care legislation of 2010 contained no provisions to eliminate unnecessary, harmful, and often lethal medical procedures.

Another step that would reduce health care costs would be to rein in the pharmaceutical companies in order to lower the cost of drugs. This will not be easy, because these companies have the largest lobby in Washington, and they spend lavishly on political campaigns.[22] Practical steps that could be taken to control prescription drug costs are discussed in the next chapter.

One more step we could take to improve the nation's health and reduce health care costs would be to provide free genetic screening for prospective parents. This should include both carrier testing (to determine whether someone carries a potentially harmful gene) and prenatal testing. Although a positive carrier test cannot foretell how severe the symptoms will be (or even if they will appear), they give parents fair warning of possible trouble ahead. No responsible parent wants to condemn a child to a life of serious health problems.

Finally, we need to reduce socioeconomic stress in society. We have to stop making people feel inferior because of their racial or economic status. When people are stressed, their glucocorticoid levels rise, making them more susceptible to cardiovascular disease, gastrointestinal disease, and psychiatric disorders. This is true even in societies with universal health care. The best way to reduce inequality in society is to ensure everyone has an adequate income and promote racial integration. Everyone benefits when societies become more egalitarian.

Unfortunately, our legislators never consider prevention in their discussions about how to control rising health care costs. Instead, they accept the nation's deteriorating health as a given and focus on how to spend more tax money on hospital care and prescription drugs. This has ensnared the nation in a deep and rapidly growing national debt. The public is never told the hard truth that they must take care of themselves. And they are offered no practical guidance on how to do this.

The first step toward changing the nation's food habits will have to be a massive educational campaign to communicate accurate dietary information to the people. Everyone should hear the good news that poor health is not inevitable. We should treat disease-producing foods the same way we do tobacco: by regulating advertising, requiring warning labels, and denying access to children. Nutritional education should be a mandatory subject from grade school through high school. Remedial nutritional education should be required for all doctors. Government subsidies for the meat, egg, and dairy industries should be halted. Government trade agencies should be prohibited from promoting food exports that cause illness.

As long as harmful foods are cheaper and more habit forming than healthful foods, even the best nutritional education will not change the dietary habits of people who make food choices based on cost and habit. Therefore, in addition to an educational campaign, it will be necessary to make the cost of harmful foods reflect what they cost society for health care and lost labor due to sickness and premature death. A simple and practical way to do this would be to impose a tax on each gram of sugar, saturated fat, sodium (in processed foods), and cholesterol. This would create a significant price differential between harmful and healthful foods, giving people an economic incentive to buy foods that keep them healthy. Denmark started to do this in 2011 by imposing a hefty tax on foods high in saturated fat. These taxes would not only improve health, but they would strengthen the economy due to a healthier labor force and less money directed to medical care. It would also raise a great deal of revenue to help pay for a universal health care program. The only people who would suffer from these measures are the ones who are killing us.

Whether we use education alone or combine it with a health-promoting tax, we can expect fierce opposition from the food and agriculture industries. They are well aware that such measures would slash the profits they make from killing hundreds of thousands of Americans every year. (Al-Qaeda, eat your heart out.) The hospital, pharmaceutical, and health insurance industries will not like dietary reform either, because their profits depend heavily on maintaining a large supply of chronically ill people.

The US needs to have a universal health care system. It will not be expensive if accompanied by an effective disease-prevention program. One of the many benefits of universal health care is that it makes it easier for employees to quit jobs they dislike. No longer would they feel chained to their employer merely to retain health insurance. And employers would benefit, too, by no longer having to pay for their employees' health insurance. Another benefit of universal health care is that it makes it easier for people who are trapped in unhappy marriages to get free. No longer would people who depend on their spouse's employer-provided health insurance feel compelled to remain in the marriage. Divorced people who pay alimony would also benefit, because they would no longer have to cover the cost of their former spouse's health insurance premiums.

The only way we are going to get genuine health care reform is if people wake up and rebel—rebel against their own unhealthy lifestyles and rebel against the greedy, parasitic corporations that rob us of our vitality and money. Our government has betrayed us, but it is ultimately our own fault for having allowed the government—and our lives—to slip out of our control.

> Gorged on edibles and imbibables, they treasure far too much. This is called thieves extravagance, the opposite of the Way.
>
> —*Tao Te Ching*[23]

Notes

1. Michel de Montaigne, *Michel de Montaigne: The Complete Works. Essays, Travel Journal, Letters*, trans. Donald M. Frame (New York: Alfred A. Knopf, 2003), 703.

2. Michael Greger, MD with Gene Stone, *How Not to Die* (New York: Flatiron Books, 2015), 9–10.

3. See the website http://www.atkinsexposed.org for accurate information about the dangerous Atkins and South Beach diets. Another good source of information on Atkins-type diets is the book *Carbophobia* by Michael Greger, MD.

4. See Caldwell B. Esselstyn Jr., MD, *Prevent and Reverse Heart Disease: The Revolutionary, Scientifically Proven, Nutrition-Based Cure* (New York: Penguin Group, 2007), 62–63.

5. The website of the US Centers for Disease Control and Prevention (https://www.cdc.gov/nchs/data/nvsr/nvsr56/nvsr56_05.pdf) acknowledges the element of arbitrariness in the creation of the list of leading causes of death: "Ranking causes of death is a popular method of presenting mortality statistics. Leading cause-of-death data have been published since 1952 (beginning with 1949 mortality data) when official tabulations ranking causes of death were first introduced (1). Users of this method of presentation should be aware of its inherent limitations. Ranking causes of death is to some

extent an arbitrary procedure. The rank order of any particular cause of death will depend on the list of causes from which selection is made, and on the rules applied in making the selection. Different cause lists and different ranking rules will typically produce different leading causes of death." An example of this is that the *Lancet* Commission on Pollution and Health in 2018 concluded that 16 percent of all deaths worldwide in 2018 were due to pollution.

6. Breast cancer has become common in Western societies. The main cause is a diet rich in fat. Secondary factors are the postponement of childbirth until the mother is in her 30s or 40s, the failure to nurse for three or four years (which causes women in Western societies have nine times more menstrual cycles than women in traditional societies), and the lack of regular exercise (female athletes have low rates of breast cancer). For most women, genetic factors are of negligible significance.

7. Alzheimer's disease is likely an expression of atherosclerosis, caused by the gradual occlusion of blood vessels in the brain by cholesterol and lipids. This reduces the supply of oxygen to the brain cells.

8. Colin T. Campbell and Thomas M. Campbell II, *The China Study* (Ben Bella Books, 2006). This book is an excellent introduction to the role of good nutrition in promoting health. Written by an outstanding scientist for laymen.

9. Lucretius, *Lucretius The Way Things Are: The De Rerum Natura of Titus Lucretius Carus*, trans. Rolfe Humphries (Bloomington, IN: Indiana University Press, 1968), book V, 188 (lines 1008–09).

10. Caldwell B. Esselstyn Jr., MD, *Prevent and Reverse Heart Disease.* For more than 35 years, Dr. Esselstyn worked as a surgeon, clinician, and researcher at the Cleveland Clinic, where he was a member of the board of governors and president of the staff. In 1991, he served as the president of the American Association of Endocrine Surgeons and organized the first national conference on the elimination and prevention of heart disease.

11. Dean Ornish, *The Spectrum: A Scientifically Proven Program to Feel Better, Live Longer, Lose Weight, Gain Health* (Ballantine Books, 2007).

12. See https://medicalxpress.com/news/2020-07-plant-based-diets-high-carbs-diabetes.html. In their book *Ever since Adam and Eve* (p. 149), Malcolm Potts and Roger Short point out that the proteins in cow milk can destroy the cells that produce insulin, creating expensive lifelong dependency on insulin injections.

13. See https://www.iarc.fr/en/media-centre/pr/2015/pdfs/pr240_E.pdf.

14. Michael Greger, MD, with Gene Stone, *How Not to Die*, 156.

15. Robert M. Sapolsky, *A Primate's Memoir* (New York: Simon & Shuster, 2001), 201–02.

16. Sharon Begley, "Sins of the Grandfathers: What Happens in Vegas Could Affect Your Offspring. How Early-Life Experiences Could Cause Permanent Changes in Sperm and Eggs," *Newsweek*, November 8, 2010.

17. Martin J. Blaser, *Missing Microbes: How the Overuse of Antibiotics Is Fueling Our Modern Plagues* (New York: Picador, 2014).

18. Ibid.

19. Sharon Begley, "This Won't Hurt a Bit," *Newsweek*, March 15, 2010.

20. B. Starfield, "Is US Health Care Really the Best in the World?" *JAMA* 284 (2000): 483–85.

21. See https://rewirenewsgroup.com/article/2010/09/14/californias-forprofit-hospitals-pushing-csections/

22. Marcia Angell, MD, *The Truth About the Drug Companies* (New York: Random House, 2004), page xx (Introduction).

23. *Tao Te Ching*, chapter 53.

Reforming the Pharmaceutical Industry

Most of the information in this chapter comes from an excellent book: *The Truth about the Drug Companies: How They Deceive Us and What to Do about It* by Marcia Angell, MD. She is the former editor in chief of the *New England Journal of Medicine* and one of the foremost authorities on drug companies and their machinations.

Americans pay the world's highest prices for prescription drugs. These high prices burden the economy by forcing everyone to pay higher federal and state taxes to support Medicare and Medicaid drug coverage and by forcing everyone who buys health insurance (individuals or their employers) to pay higher premiums. The uninsured suffer the most, because high drug prices often result in financial ruin or death due to inability to afford drugs.

Contrary to popular assumption, drug companies rarely discover and develop their own drugs. Instead, they troll around universities and small biomedical companies looking for promising new discoveries and buy up the rights. US taxpayers fund most discoveries of new drugs through grants made by the National Institutes of Health (NIH) to university researchers. However, we Americans are not allowed to enjoy the fruits of our investments. Instead, Congress, through two laws called Bayh-Dole and Stevenson-Wydler, has decided that the fruits should go to pharmaceutical companies. Moreover, once the pharmaceutical companies acquire these drugs, they are allowed to jack up the price as high as the market will bear. And lest anyone try to acquire these drugs more cheaply from another country via the internet, Congress has cut off that option. Likewise, Congress has forbidden the administrators of Medicare from trying to place reasonable restraints on drug prices.

The anti-AIDS drug AZT provides a good illustration of how Bayh-Dole and Stevenson-Wydler work to maintain high drug prices. Developed in 1964 by the Michigan Cancer Foundation, AZT proved unsuccessful as a cancer treatment. However, a German laboratory subsequently discovered it was effective against viruses in mice. This prompted the pharmaceutical company Burroughs Wellcome to acquire rights to it as a possible herpes treatment. Meanwhile, researchers at Duke University, operating under grants from the National Cancer Institute (part of NIH), discovered that AZT killed the AIDS virus in test tubes. Excited by their discovery, they set up a clinical trial, which confirmed the effectiveness of AZT in humans. Upon hearing this good news, Burroughs Wellcome promptly patented AZT as an AIDS treatment. The company then proceeded to charge those dying of AIDS an outrageous $10,000 for a one-year supply.

Besides taking advantage of drugs developed by taxpayer-funded research, drug companies further exploit the public by finding ways to keep their patent monopolies in effect long after the patent expires. One way they do this is by obtaining a patent on a new drug that is nearly identical to the old one—both chemically and in its mode of action. This is what AstraZeneca did with Prilosec. Just before Prilosec's patent was due to expire, they introduced Nexium, a nearly identical chemical. They then spent $500 million promoting Nexium. Many users switched from Prilosec to Nexium without even knowing that Prilosec had become available as a generic drug at a far lower price. This same approach to extending monopoly protection was used by Schering-Plough with its blockbuster drug Claritin. Just before the patent expired, Schering-Plough patented a drug called Clarinex, which was nothing more than the chemical Claritin breaks down to in the body. They then launched a major advertising campaign to persuade allergy sufferers to buy the expensive Clarinex instead of the equally effective but inexpensive generic version of Claritin.

There are many problems with the way the Food and Drug Administration (FDA) regulates prescription drugs. For example, in order to obtain drug approval, drug companies need only convince the FDA that their drug is more effective than a placebo, even if only slightly. In other words, they don't have to compare the effectiveness of the new drug against existing drugs. This means that doctors have no way of knowing which drugs in the US pharmacopoeia are most effective against specific ailments.

The FDA allows drug companies to fund their own clinical trials. These are mostly carried out by contract research organizations (CROs) or by medical schools and teaching hospitals that receive their funds from the drug companies. This is equivalent to letting students grade their own papers. Drug companies are well known to suppress unfavorable results, spin whatever results they do publish, conduct their trials by comparing their drugs to excessively small doses of other drugs, and publish only favorable portions of their studies. They also prefer to use young subjects for their trials because they know this will reduce the incidence of side effects. Especially disturbing is the refusal of drug companies to publish research results that show the tested drugs to be ineffective or even harmful. Although the National Institutes of Health has a website called ClinicalTrials.gov, which "is a registry and results database of publicly and privately supported clinical studies of human participants conducted around the world," there is *no requirement* that negative results be published.

Instead of letting drug companies test their own drugs, the FDA should conduct all clinical drug trials itself. The FDA should start by retesting our existing pharmacopoeia to find out which drugs are most effective against specific ailments. Once we know which drugs are best, the FDA should purchase the rights to those drugs at a fair price. The FDA should then solicit bids from the drug companies to manufacture the drugs. The drug companies would still be allowed to conduct their own independent research, and if they discover something useful, the government would test it and buy the rights to it at a fair price. Putting the FDA in charge of drug screening and purchasing would have these benefits:

- Reduce drug prices.
- Reduce mortality rates.
- Reduce employee out-of-pocket health insurance costs and paycheck deductions for Medicare and Medicaid.
- We would no longer suffer shortages of drugs that drug companies don't find profitable enough to manufacture, such as orphan drugs, antivenins, and vaccines for children. Whenever a drug was in short supply, the FDA would simply solicit bids to manufacture it.
- The public would no longer be bombarded by television drug advertising. Such direct-to-consumer advertising is

already banned in most countries. Doctors would pre-scribe drugs based on the best scientific evidence, not because ill-informed patients demanded a drug promot-ed on TV.

- Drug advertisements in medical journals would cease, decreasing the likelihood that doctors will be swayed by advertising instead of sound scientific studies.

- Personal sales pitches to doctors by drug-company sales representatives would end. There are currently some 90,000 drug-company sales representatives in the US—an amazing figure—and they visit the average doctor several times a week.[1]

- Drug companies would no longer offer doctors free drug samples. The purpose of these samples (whose costs, of course, are passed on to consumers) is to habituate pa-tients to the company's latest and most expensive offer-ings.

- Drug companies would no longer sponsor "educational" seminars for doctors in which doctors are given continu-ing education credits. The real purpose of these seminars is to promote the sponsor's drugs.

- Drug companies would no longer pay for doctors' mem-berships in professional medical societies. If doctors want to join a medical society, they would have to pay the fees themselves. Imagine that!

- Drug companies would no longer set up patient advo-cacy groups as a surreptitious stratagem to increase de-mand for specific drugs. Nor would they be allowed to dream up and promote new "illnesses" for which their drugs just happen to be the "cure."

These reforms will not be welcomed by everyone. The price of pharmaceutical stock will plummet, Big Pharma's nearly 90,000 sales representatives will lose their jobs, and many doctors are sure to lament the loss of drug company largesse: no more kickbacks and no more junkets masquerading as "educational seminars."[2] Doctors would also have to pay for their own memberships in medical societies and pay more for their continuing education credits. The government should continue to fund research through the National Institutes of Health. However, Bayh-Dole and Stevenson-Wydler should be repealed, so that taxpayers can retain control of all drugs discovered through research they fund. We need to stop handing over promising new drugs to drug companies so they can sell them back to us at exorbitant prices. The NIH also needs to prohibit conflicts of interest among its employees, such as consultant relationships with pharmaceutical companies. Prior to 1995 it had a ban on such relationships.

The key to enacting all these reforms (and many others) will be to establish public financing of political campaigns. There is no other way for the people to recapture their government and make it begin to serve their interests. Right now, the pharmaceutical lobby is the most powerful in Washington, and it controls nearly every member of Congress. Until we can recapture the government, the best we can do is try to persuade our elected "representatives" to pay attention to our proposals and refuse to reelect them when they don't.

Notes

1. Sales representatives can be quite sophisticated in their methods. One technique is to buy information about doctors' prescribing habits from prescription-tracking companies in order to learn exactly which drugs a particular doctor is prescribing. This allows the sales rep to tailor his sales pitch to each doctor. (*The Truth About the Drug Companies*, pages 129–30.)

2. The prime purpose of these seminars (which are often held at fancy resorts) is to persuade doctors to prescribe drugs for "off-label" ailments (that is, ailments for which the drugs have no proven value). Although it is illegal for drug companies to market drugs for off-label ailments, it is not illegal for them to "educate" doctors about other potential uses for a drug; and doctors are legally allowed to prescribe a drug for *any* use. Doctors appreciate these seminars not only because they are feted but because they obtain credits for continuing medical education, which they need to maintain their licenses.

Making Hospitals More Conducive to Healing

Anyone who has spent even one day in a typical American hospital knows that these institutions are operated not for the benefit of the patients but for the profit of the administration and the convenience of the staff. Patients often feel like prisoners, and for all practical purposes they are.[1] Patients ultimately pay the salaries of doctors, nurses, and hospital administrators, so they should be treated with due respect. There should be a patient's Bill of Rights prominently displayed in every room, and patients should be read these rights in their native language upon admittance. Patients should be free to leave the hospital at any time. If a patient wants an intravenous line or a catheter removed, the staff should be required to comply immediately.

The food served at most hospitals is generally high in saturated fat, cholesterol, and sugar—the very fare that brought most of the patients to the hospital in the first place. Hospitals should be serving meals that are both appetizing and healthful. For most patients, this would be their first exposure to a good diet and would show them that healthful foods can be delicious.

The decor of American hospital rooms is generally drab and depressing. In Catholic hospitals there may be nothing more on the wall than a statuette of Jesus being crucified. Imagine lying with an intravenous line in each arm and a catheter in your urethra and looking up at a crucifixion all day! Hospital rooms should have big windows that look out on a natural landscape, allowing patients to have contact with nature and fresh air. Patients should also have access to good movies, books, magazines, music, and the internet.

The standard hospital "gown" is one of the ugliest and most degrading garments ever conceived. Hospital attire should be attractive, dignified, and comfortable. Whatever attire the patients wear is what the doctors, nurses, and administrators should wear. Wheelchairs, too, should be redesigned to be stylish.

Catheters should be inserted only when absolutely necessary. Currently, they are often used for the mere convenience of the staff, who are thereby saved the trouble of escorting a patient to the toilet. The result is discomfort and bondage for the patient and, all too often, infection.[2]

Hospitals should have quiet rooms where patients can meditate or gather to take ayahuasca, peyote, or psilocybin mushrooms with experienced guides (or alone if they prefer). Mankind has recognized for thousands of years that these plants act powerfully to heal body and spirit. Modern research has shown that they can also help the terminally ill come to terms with death.[3]

Touch is a powerful form of healing that is completely ignored in today's hospitals. Holding hands, stroking a head, massaging feet, are calming and reduce pain. A relaxing massage by an attractive nurse would go a long way toward restoring the will to live in many patients.

The terminally ill should always have the option of ending their lives with dignity whenever they wish, or of designating someone they trust to make this decision for them when they can no longer make it for themselves. This would not violate the "do no harm" provision of the Hippocratic oath, because it would be a far greater harm *not* to provide this service.

> Better by far to blast out your brains with a shotgun, to dive off a cliff, to plunge into the river, than inflict upon yourself and upon those who care for you this piecemeal, inchworm, smelly, messy, and rotten and horribly prolonged dying. It's unfair not only to the victim, but also to everyone who is forced to attend it. Worst of all, it is a sort of insult to human dignity, to the human spirit.
>
> —Edward Abbey[4]

There is moral symmetry between determining when to give birth and when to die. Both are uniquely human capacities derived from our ability to foresee. Just as contraception enables us to avoid inflicting famine, pestilence, and war on posterity, so does determining the time of our demise spare our families and friends the expense and unpleasantness of our final physical and mental decay and collapse.

Hospitals should halt expensive interventions that merely prolong the lives of the moribund for a few days. It wastes money that should instead be spent on disease prevention and the treatment of patients with hopeful prospects.

There are many other hospital reforms that should be carried out, and they will be as soon as hospitals are put in the service of healing instead of profit.

Notes

1. A case in point is what Joseph Wheeler alleges happened to him at Prince George's Hospital in Cheverly, Maryland. On June 23, 2010, he was brought to the hospital in an ambulance after suffering injuries in an automobile crash. The hospital put a bracelet on him that was intended for another patient—one who was scheduled for surgery to remove a tumor. When Wheeler learned he was going to be wrongly operated on, he tried to leave the hospital, but was prevented by security guards who insulted and beat him. Wheeler and his wife sued the hospital for assault and battery, false imprisonment, and infliction of emotional distress. See www. courthousenews.com /2010/08/25/29858.htm.

2. See http://seniorjournal.com/NEWS/Alerts/6-05-22-OlderPatientsBeing.htm.

3. R. R. Griffiths, W. A. Richards, M. W. Johnson, U. D. McCann, and R. Jesse, "Mystical-Type Experiences Occasioned by Psilocybin Mediate the Attribution of Personal Meaning and Spiritual Significance 14 Months Later," *The Journal of Psychopharmacology* 22, no. 6 (2008), 621–32. This study was conducted at Johns Hopkins University. For a fascinating firsthand account of long-term remission of terminal cancer following experiences with ayahuasca, see Professor Donald Topping's story "Making Friends with Cancer and Ayahuasca" in *Hallucinogens: A Reader*, ed. Charles S. Grob (Jeremy P. Tarcher, 2002), and a follow-up at https://maps. org/news-letters/v08n3/08322top.html.

4. Edward Abbey, *Confessions of a Barbarian* (Boulder: Johnson Books, 1994), 190.

Part 7:

Ethical Reforms

Developing Worthwhile
Values and Pursuits

A steady state economy (one based on a stable population) would allow us to improve the art of living, as John Stuart Mill observed:

> It is scarcely necessary to remark that a stationary condition of capital and population implies no stationary state of human improvement. There would be as much scope as ever for all kinds of mental culture, and moral and cultural progress; as much room for improving the Art of Living, and much more likelihood of its being improved, when minds ceased to be engrossed by the art of getting on. Even the industrial arts might be as earnestly and as successfully cultivated, with this sole difference, that instead of serving no purpose but the increase of wealth, industrial improvements would produce their legitimate effect, that of abridging labour. Hitherto it is questionable if all the mechanical inventions yet made have lightened the day's toil of any human being. They have enabled a greater population to live the same life of drudgery and imprisonment, and an increased number of manufacturers and others to

make fortunes. They have increased the comforts of the middle classes. But they have not yet begun to effect those great changes in human destiny, which it is in their nature and in their futurity to accomplish. Only when, in addition to just institutions, the increase of mankind shall be under the deliberate guidance of judicious foresight, can the conquests made from the powers of nature by the intellect and energy of scientific discoverers, become the common property of the species, and the means of improving and elevating the universal lot.[1]

When Mill says "there would be as much scope as ever for all kinds of mental culture, and moral and cultural progress," he calls attention to the fact that knowledge—unlike material resources and populations—can potentially grow forever. Knowledge is not limited by the physical laws that constrain matter and energy.

Unfortunately, in modern societies, there doesn't seem to be much enthusiasm for mental culture, nor for moral and cultural progress. Instead, most people spend their lives trying to win respect by acquiring and displaying material wealth. In the sexual arena, this takes the form of men competing to accumulate enough wealth to win over women. Women, in turn, compete against other women to win over a man of suitable income and status. To this end, they use clothes, cosmetics, perfumes, and even plastic surgery. And they go to the colleges, churches, clubs, bars, and internet dating sites where they feel they are most likely to meet an appropriate man. This competition turns everyone into losers, because it makes everyone feel insecure, envious, anxious, self-critical, angry, lonely, and sad. No matter how much we have, others will always have more. This situation is made worse by advertising, which tells us we are inadequate unless we buy the advertised product. Wendell Berry observed that one of the strangest claims of the cult of competition "is that the result of competition is inevitably good for everybody, that altruistic ends may be met by a system without altruistic motives or altruistic means."[2] We need to recognize that competition works against democracy, community, and mutual care.

Some people have begun to recognize that they already have enough. They are discovering, as Thoreau did, that "a man is rich in proportion to the number of things which he can afford to let alone."[3] They are coming to realize that what matters in life is to have a healthy mind in a healthy body and to do beneficial work while enjoying the love and support of family and friends.

> The greatest calamity is not knowing sufficiency.
> —*Tao Te Ching*[4]

Defenders of consumerism point out that it encourages technological advances by providing a market for them. So what? As economist Leopold Kohr pointed out, the creation of new necessities should not be taken as a sign of progress.[5] Our goal should a happier, wiser society, not technology for technology's sake. Most of today's products are redundant, frivolous, and nonrecyclable. Their production creates pollution, wastes resources, and wastes the labor of those who make them and those who buy them. Even products that are useful are typically designed to wear out quickly. Economist Kenneth Boulding pointed out that products that lack durability significantly reduce a country's gross national product.[6] He added the following:

> Any discovery which renders consumption less necessary to the pursuit of living is as much an economic gain as a discovery which improves our skills of production.[7]

Instead of consumerism, we need to temper our wants and restrict our purchases to products that are useful, durable, repairable, recyclable, and don't degrade the environment.

One of the pioneers working to make products that are useful, durable, recyclable, and less expensive is Marcin Jakubowski. He is developing open-source designs for inexpensive, highly useful machines such as farm tractors.[8] This kind of open-source technology may eventually result in the de-monopolizing of production and a radically different kind of economy. As people reduce their consumption of goods and services, they will simultaneously reduce the amount of time they need to spend working. With more free time, people will be able to pursue worthwhile goals that matter to them.

> Let us define the wealthy man as he who has everything he desires. How to reach that happy condition. Two ways…

> 1. Through money: work, sweat, scheme, grovel, cheat, lie, betray to acquire it. But there's no guarantee you'll succeed. Ninety-nine chances out of a hundred, you'll fail. Or…

2. Do without: reduce your needs to the minimum required for a healthy life. Get by on part-time work. Enjoy the leisure of the leisure class. That's the easy way to become rich, and anyone can do it; the success rate is one-hundred percent.

—Edward Abbey[9]

As computers and robots increasingly take over jobs in both the manufacturing and service sectors, and as ever more jobs are off-shored, an ever-growing proportion of the population will become permanently unemployed. Unemployment has traditionally been accompanied by increased alcoholism, domestic violence, suicide, stress-induced disease, and civil unrest. To avoid these ugly consequences, the unemployed will need new kinds of employment that will utilize their talents for the benefit of themselves and society. They will also need a universal basic income as a backstop against poverty.

Television and its internet equivalents displace reading, communication, and physical activity and thereby contribute to ignorance, loneliness, and physical debility. Edward Abbey likened TV to "reality reduced to flickering shadows on an illuminated screen. Every home a platonic cave."[10] Even the inventor of TV, Philo T. Farnsworth, came to despise it. According to Farnsworth's son, "he felt he had created kind of a monster, a way for people to waste a lot of their lives."[11] Without TV, people would have healthier minds and bodies, would waste less money, and would have more time to be in nature, to gain and share worthwhile knowledge, and to spend time with family and friends.

Next to selfishness, the principal cause which makes life unsatisfactory is want of mental cultivation. A cultivated mind—I do not mean that of a philosopher, but any mind to which the fountains of knowledge have been opened, and which has been taught, in any tolerable degree, to exercise its faculties—finds sources of inexhaustible interest in all that surrounds it; in the objects of nature; the achievements of art, the imaginations of poetry, the incidents of history, the ways of mankind, past and present, and their prospects in the future.

—John Stuart Mill[12]

A mind that examines "the ways of mankind, past and present" is a mind less likely to identify too closely with groups:

A man who knows that there have been many cultures, and that each culture claims to be the best and truest of all, will find it hard to take too seriously the boastings and dogmatizings of his own tradition.

—Aldous Huxley[13]

We cling to group identities because we crave social acceptance. This begins when we are infants, striving to get our mother's favorable attention. It continues throughout life as we try to meet society's expectations in order to get a job, win and keep a mate, avoid ostracism, and stave off premature death. Our dread of rejection and our longing for praise is what enables authority to manipulate us.

The only antidote to our fear of social rejection is to develop confidence in our own judgment and intuition, and to learn to accept death as Socrates did. Instead of seeking praise and fleeing censure, we should develop a love of learning, a love for clear thought and speech, and a determination to treat others (including other species) with respect. Much of the behavior that society rewards with praise is foolish and destructive. We should never confuse praise with love. Praise is conditional and manipulative, but love is unconditional and accepting.

A cultivated mind is also a spiritual mind. Economist Kenneth Boulding suggested that as mankind matures, "the end results may well be a society specializing in spiritual experiences of a quality which we now realize only in rare moments of intuition."[14] In the Western world, interest in spiritual experience grew rapidly in the 1960s as young people began reading about Eastern religions and learned of the existence of higher consciousness. Some even took up meditation, aspiring to follow the Buddha and attain enlightenment. But it was only when LSD became widely available in the mid-1960s that large numbers of people were finally able to discover for themselves what higher consciousness was about. These explorations seemed strange and frightening to those in power, and they reacted by outlawing psychedelics. They even outlawed nearly all scientific research on them—one of the few bans on scientific inquiry since the Inquisition. In recent years, a few governments have relaxed enough to allow new scientific studies of psychedelics. What is clear from both the new research and that conducted prior to 1966 (when LSD was outlawed) is that psychedelics are not only powerful tools for exploring consciousness, but they can help people come to terms with terminal illness,[15] can help to cure addictions, can relieve depression, can help to heal posttraumatic stress syndrome, can effectively treat cluster headaches, and can stimulate creativity.[16] By helping to heal hearts and minds, they help to heal societies.

> The state of the world now reflects the state of our minds. What we call our global "problems" are actually "symptoms" of our individual and collective mind states. If we are to be effective in transforming the crises we face, we must work, not only to reduce overpopulation and feed the hungry, but to reverse the psychological states, limitations, and perceptual distortions that allowed us to create these things in the first place. We are going to have to work in both arenas, in the world and in ourselves, if we are to effect a healing.
>
> —Roger Walsh[17]

Gerald Heard, one of the 20th century's preeminent sages, recognized that psychological change is an evolutionary imperative for humans:

> If then, the whole purpose of our evolution and existence is the advancement into a higher and more extended consciousness, if we have now reached a stage when that evolution must, to achieve its purpose, for the preservation of civilization and for the development of ourselves, be intentionally pursued, it is clear that nothing else so much matters as the attaining to such a power. If we fail to advance—

achieve this enlargement of consciousness—then we are ourselves frustrated. Our descent to Avernus may go by one path or another, slow or precipitous, but it will reach the same goal—the destruction of a species which balked at undertaking further evolution.[18]

Psychedelics will continue to play a key role in mankind's social and spiritual advancement. But it takes time and effort to integrate the knowledge they impart. They are not panaceas, and they are definitely not for everybody.

> I share the belief of many of my contemporaries that the spiritual crisis pervading all spheres of Western industrial society can be remedied only by a change in our worldview. We shall have to shift from the materialistic, dualistic belief that people and their environment are separate, toward a new consciousness of an all-encompassing reality, which embraces the experiencing ego, a reality in which people feel their oneness with inanimate nature and all of creation.
>
> —Albert Hofmann, discoverer of LSD[19]

Notes

1. John Stuart Mill, *Principles of Political Economy* (New York: Prometheus Books, 2004), book 4, chapter 6, 692. Originally published in 1848.

2. Wendell Berry, *What Matters? Economics for a Renewed Commonwealth* (Berkeley: Counterpoint, 2010), 94–96.

3. Henry David Thoreau, *Walden, or Life in the Woods* (New York and Scarborough, Ontario: New American Library, 1960, 1980), 60.

4. Lao Tzu. *Tao Te Ching: A New Translation*, trans. Sam Hamill (Boston: Shambhala Publications Inc., 2005), chapter 46.

5. Leopold Kohr, *The Breakdown of Nations* (Devon: Green Books Ltd., 2012), 148–150. Originally published in 1957.

6. Kenneth Boulding, *Three Faces of Power* (Newberry Park, CA: Sage Publications, 1989), 193.

7. Kenneth Boulding, "The Consumption Concept in Economic Theory," *American Economic Review*, May 1945, p. 2.

8. See http://opensourceecology.org and http://openfarmtech.org/ weblog.

9. Edward Abbey, *Confessions of a Barbarian: Selections from the Journals of Edward Abbey* (Boulder: Johnson Books, 1994), 35. Thoreau felt as Abbey did: "Those slight labors which afford me a livelihood, and by which it is allowed that I am to some extent serviceable to my contemporaries, are as yet commonly a pleasure to me, and I am not often reminded that they are a necessity. So far I am successful. But I foresee that if my wants should be much increased, the labor required to supply them would become a drudgery. If I should sell both my forenoons and afternoons to society, as most appear to do, I am sure that for me there would be nothing left worth living for. I trust that I shall never thus sell my birthright for a mess of pottage" (*Life Without Principle*, 1854).

10. Edward Abbey, *Confessions of a Barbarian: Selections from the Journals of Edward Abbey* (Boulder: Johnson Books, 1994), 147. Readers of *The Monkey Wrench Gang* will recall the delightful scene in which Doc Sarvis smashes the screen of his television.

11. See www.byhigh.org/History/Farnsworth/PhiloT1924.html.

12. John Stuart Mill, *Utilitarianism* (1863), chapter 2.

13. Aldous Huxley, "Culture and the Individual" (1963) in *Moksha: Aldous Huxley's Classic Writings on Psychedelics and the Visionary Experience*, ed. Michael Horowitz and Cynthia Palmer (Rochester, Vermont: Park Street Press, 1999), 249.

14. Kenneth Boulding, *The Meaning of the 20th Century: The Great Transition* (New York: Harper Colophon, 1964), 155.

15. R. R. Griffiths, W. A. Richards, M. W. Johnson, U. D. McCann, and R. Jesse. 2008. "Mystical-Type Experiences Occasioned by Psilocybin Mediate the Attribution of Personal Meaning and Spiritual Significance 14 Months later" *Journal of Psychopharmacology* 22, no. 6: 621–32. This study was conducted at Johns Hopkins University.

16. See *Psychedelic Medicine* (vols. 1 and 2), edited by Michael J. Winkelman and Thomas B. Roberts for the best review of the therapeutic benefits of psychedelics. For more on the link between psychedelics and creativity, see *LSD Spirituality and the Creative Process* by Marlene Dobkin de Rios and Oscar Janiger.

17. Roger Walsh, "Consciousness and Asian Traditions: An Evolutionary Perspective" in *Psychoactive Sacramentals: Essays on Entheogens and Religion*, edited by Thomas B. Roberts (San Francisco: Council on Spiritual Practices, 2001), 183.

18. Gerald Heard, *Pain, Sex, and Time: A New Outlook on Evolution and the Future of Man* (Rhinebeck, NY: Monkfish Book Publishing Company, 1939), 169–70.

19. Albert Hofmann, *LSD: My Problem Child; Reflections on Sacred Drugs, Mysticism, and Science* (Multidisciplinary Association for Psychedelic Studies, 2009), 31.

Reconciliation with Other Animals

As people advance in spiritual and ethical awareness, they will abandon the practice of enslaving, killing, and eating other animals. This
practice is bad for us and much worse for the animals. For us, the effects have been infectious and chronic diseases (such as COVID-19
and heart disease), air and water pollution, global warming, aquifer
depletion, soil erosion, reduced wealth, and spiritual coarsening.[1]
We consume meat out of culturally conditioned habit and for the
fleeting pleasure of the taste. We ignore, or try to ignore, that other
animals love their lives as much as we love ours and feel pain just as
we do. Taking another's life in order to briefly savor its flesh is a grave
moral wrong. There is no ethical system in the world that could condone such a mindless and cruel practice.

The eating of animals not only deprives them of life, liberty, and
the pursuit of happiness, but does so in a way that inflicts gratuitous
suffering. This suffering has greatly intensified in recent decades as
animal exploitation has become industrialized. In today's factory
farms, egg-laying hens are jammed together in tiny cages that create
intense stress and injuries; pregnant and nursing pigs are confined
to cages with concrete floors that are so confining they cannot even
turn around; and calves are torn away from their mothers at birth
to be confined to "veal crates" where they are fed a diet intended to
make them anemic (pale veal is considered desirable). Pigs have their

tails cut off with no anesthetic, chickens have the outer third of their highly sensitive beaks seared off, castrations are performed without anesthetics (causing acute stress and prolonged pain), no medical care is provided to the sick and injured (who are numberless)—the litany of horrors is appalling. Edward Abbey commented on what these animals endure and the implications:

> Perhaps my hero Solzhenitsyn would scorn my saying so but I am tempted to believe that the systematic cruelty inflicted upon animals trapped in our food and research apparatus is comparable—for who can measure the aggregates of pain, the sum of suffering?—to the agony that contemporary despotisms have exacted from human beings caught in their archipelagos of tyranny. Not merely comparable but analogous. Not merely analogous but causally connected. Contempt for animal life leads to contempt for human life.[2]

If mankind eventually advances far enough to adopt a plant-based diet, it will also need to strictly regulate birth rates. That is because a plant-based diet can support far more people on a given amount of land than a meat-based diet.[3] Without strict population control, the adoption of a plant-based diet would lead to runaway population growth.

On Hunting and Fishing

The proponents of hunting and fishing call these activities "sports." But unlike true sports, where participation is always voluntary and the contestants are evenly matched, the targets of hunters and sport fishers are neither volunteers nor equal opponents. To hunt and fish means to deceive, betray, terrorize, maim, and kill. These behaviors violate the precepts of every ethical system but one: that of the perpetrators. Those who think that hunting and fishing are sports should try playing on the other team.

Hunters and fishers violate not only animals, but they deny the nonhunting and nonfishing public (who are the majority) the opportunity to enjoy the outdoors in peace, free from flying arrows, gunshots, trails of blood, and the desperate flopping of dying fish. They also deprive the public of the pleasure of observing fearless animals close by.

> They belonged to me, as much as to anyone, when they were alive, but it was considered of more importance that Mrs. X should taste the flavor of them dead than that I should enjoy the beauty of them alive.
>
> —Henry Thoreau[4]

Defenders of hunting argue that hunters play a valuable role by keeping herbivore populations in check. That is true, but it is only true because hunters have killed off the natural predators. Moreover, human hunters do a poor job of it, because they preferentially kill prime specimens (the genetically fittest), whereas natural predators preferentially kill the old, the sick, and the excess young.

The motivations of sport hunters are various. Some like to pretend they are living off the land like frontiersmen. Others hope to slay an animal of record size so they can boast of it. Some wish to impress their companions with their marksmanship. Others think that by killing the defenseless they prove their masculinity. Yet others are lured by the intense excitement they feel upon extinguishing another's life. (Human executioners report the same excitement—followed by depression.)[5] Regardless of motive, the act of hunting always requires that hunters suppress their capacity for empathy—just as actors must do when called upon to play villains. Hunting builds barriers, hardens egos, and blocks spiritual development.

> Evil is the accentuation of division; good, whatever
> makes for unity with other lives and other beings.
> —Aldous Huxley[6]

Some have suggested that the hunting phase of mankind's past was a golden age and that the advent of agriculture opened Pandora's box, creating overpopulation, slavery, cities, pollution, and epidemics.[7] That view is naive: paleohunters wrought enormous environmental destruction. Wherever they went, they sought out and slaughtered the largest animals. Among their victims were mastodons, mammoths, giant beavers, woolly rhinos, giant tortoises, glyptodonts, diprotodonts, giant kangaroos, horned turtles, ground sloths, elephant birds, and moas—all now extinct.[8] Like Easter Island's colonizers, paleohunters had no concept of conservation. To them, the world was an inexhaustible commons, and they adhered to that ignorant belief despite wiping out one species after another.

The first human to be infected with the HIV virus was an African hunter who, while butchering a chimpanzee, acquired the virus through blood contact. Due to that hunter's violence, 37 million people now carry the HIV virus; and more than 35 million have died of AIDS, and the death toll rises by 5,500 each day. Fifteen million children have lost one or both of their parents to AIDS, and countless people have lost their friends and teachers. No one could have anticipated that killing a chimpanzee would unleash these terrible consequences, but if humans had respected the lives of other animals, the AIDS epidemic would never have occurred.

The deadly Ebola virus is another consequence of hunting. This virus normally lives in fruit bats (which are immune to it). Chimpanzees originally acquired the disease by eating fruit bats, and humans got it by eating infected chimpanzees. The disease then spread among humans through contact with bodily fluids.[9]

SARS-CoV-2 (the COVID-19 virus) is another fruit bat virus. Most scientists believe it spread from bats to an unidentified wild mammal in the Wuhan animal market before jumping to humans. Live-animal markets are notorious spreaders of new diseases. The immense suffering caused by COVID-19 once again demonstrates how cruelty inflicted on animals can rebound against us.

Even pubic lice ("crabs") are a consequence of hunting. DNA evidence indicates that the original host of pubic lice was gorillas.[10] The lice must have passed to humans when hunters killed and butchered a gorilla.

Fishing deserves special comment. By means of cruel experiments, scientists have established that fish are highly sensitive to pain and that morphine can numb their pain. Not only is the experience of being caught on a hook excruciating for the fish, but the resulting stress and injuries cause between 30 and 70 percent of released fish to die of their injuries in a matter of days.[11] In fact, catch and release may result in higher rates of fish mortality than if every sport fisher were to catch and keep the legal limit. If fish could scream, fishing would have been banned since the advent of conscience. We are descended from fish. It is time to wake up.[12]

Sportsmen are adept at rationalization. They have to be in order to preserve their self-respect. If they faced the truth, they would have to change, and change requires courage. It takes even more courage to explain the change to our friends and relatives, who might not like what it implies about themselves. Yet it is only by changing that we can advance—as individuals and as a people.

Our primate ancestors took up hunting (meat-eating) because their populations grew too large for the local supply of seeds, nuts, fruits and vegetables. One consequence of thousands of years of eating animals is that humans, unlike other primates, can no longer absorb sufficient vitamin B_{12} from our intestinal microbiome. Instead, we must obtain B_{12} from animal foods or (if we are vegans) from golden algae or commercial foods supplemented with B_{12}.

Once humans began hunting animals, it was but a small step to begin hunting fellow humans. In their book *Sex and War,* Malcolm Potts and Thomas Hayden point out that traditional hunting practices share many features with homicidal team aggression:

- Cooperation within a small group of related males.
- Prey is taken by ambush.
- The struggle is usually brief.
- The risk of injury to the aggressors is small (provided group members cooperate).

- There is no sympathy for the victim.
- Females (if present) become excited (because the males share the fresh kill with them).
- There is an immediate benefit for the aggressors. Hunters obtain more food and prestige, and warriors get more territory (food resources) and more females for sex.

Human violence is a product of natural selection. Nature bequeathed us the capacity to switch off empathy and become horrifically cruel. This ability to switch off empathy is the true root of evil. We will only end evil when we eliminate this switch through genetic engineering.

Cruelty in Lab Coats

Hundreds of millions of nonhuman animals are tortured and killed in laboratories every year. The defenders of this violence argue that once one has embraced a noble goal (such as finding a cure for a disease), the infliction of suffering and death on others to further that goal is morally justifiable. This argument is called the "higher good" or "the end justifies the means." However, it is an argument based on a misunderstanding of the way in which ends and means relate to time. Means exist in the present; ends pertain to the future. Since we can only live in the present, means alone have reality and moral significance. That is why we must do the right thing *now* and never pretend that doing the wrong thing will somehow get us to the right thing. (As if the universe had moral wormholes!) We can no more make ourselves healthy by killing animals than we can cure Peter by killing Paul. Beneath the surface, humans and other animals are the same. That is true both scientifically and spiritually.[13]

Some have tried to defend animal research by arguing that useful information has been obtained from these cruelties. That is undeniably true, but the most useful information for human health comes from studying humans. Unfortunately, the mistaken beliefs and distorted values of animal experimenters are deeply embedded and hard to uproot. Like all who kill in vain, they need a spiritual breakthrough to help them see the vanity of their ambition.

> …the motive has nothing to do with the morality of the action, though much with the worth of the agent. He who saves a fellow creature from drowning does what is morally right, whether his motive be duty or the hope of being paid for his trouble; he who betrays the friend that trusts him, is guilty of a crime, even if his object be to serve another friend to whom he is under greater obligations.
>
> —John Stuart Mill[14]

Those who experiment on animals do so to enhance their professional status and make money, while asserting that animal sacrifices benefit the public. In so doing they reenact the role of the animal-sacrificing priests of ancient times, who likewise claimed that only by killing animals could the well-being of society be secured.

> I saw the laboratory animals: throat-bandaged dogs cowering in cages, still obsessed with the pitiful

Love that dogs feel, longing to lick the hand of their
devil; and the sick
monkeys, dying rats, all sacrificed
To human inquisitiveness, pedantry and vanity, or at
best the hope
Of helping hopeless invalids live long and hopelessly.
—Robinson Jeffers[15]

Exhibiting Captives for Entertainment

What is an eagle in captivity!—screaming in a court-
yard! I am not the wiser respecting eagles for having
seen one there.

—Henry Thoreau[16]

The United States has some 2,000 internment camps called zoos, marine parks, and aquariums. The inmates of these prisons are cut off from their natural habitats and often denied their natural social life. They are held captive so that money can be made by exhibiting them to a gawking public. All zoos, marine parks, and aquariums should be closed down as soon as their inmates can be safely released to the wild or die off. Only facilities for breeding endangered wildlife, for treating and rehabilitating injured wildlife, and for housing animals that cannot live on their own should be retained. Unfortunately, as ever more animals are driven toward extinction, survival in captivity will increasingly become the only option to keep them going until human populations are finally reduced and people learn to live nonviolently with their fellow beings.

Signs of Hope

Despite all the horrors still being inflicted on animals, more enlightened attitudes are beginning to emerge. The proportion of the population that hunts and fishes is declining (now about 5 percent in the US), and laws are beginning to be enacted against the worst abuses of factory farming. In the more advanced societies, nonviolent diets are making headway. This was anticipated by Thoreau:

> Whatever my own practice may be, I have no doubt that it is a part of the destiny of the human race, in its gradual improvement, to leave off eating animals, as surely as the savage tribes have left off eating each other when they came in contact with the more civilized.[17]

Notes

1. For a good explanation of the negative environmental impacts of meat eating, see the United Nations report "Livestock's Long Shadow" (2007). Available at https:// ftp.fao.org/docrep/fao/010/ a0701e/a0701e00.pdf.

2. Edward Abbey, "A Writer's Credo" in *One Life at a Time, Please* (New York: Henry Holt and Company, 1987), 170–71.

3. Not only can a plant-based diet support far more people than a diet heavy on animal products, but certain plants (such as potatoes) can support far more people than cereal grains. Malthus pointed out that a population whose diet is based on wheat will not suffer unduly if the wheat harvest fails, because they can always fall back on potatoes and grains like oats, barley, and rye. But an overpopulated society

that depends exclusively on potatoes has nothing left to fall back on. That was the fate of the Irish in the 1840s.

4. Henry David Thoreau, *Journal*, August 16, 1858. He was referring to a pair of wild ducks which he had enjoyed observing, but which were shot by a hunter, who gave them to a woman.

5. Michael Daly, "I Committed Murder: For the Anonymous Executioners of Death Row, the 'High' of Pulling the Lever Is Often Followed by a Lifetime of Doubt," *Newsweek*, October 3, 2011, 42–45.

6. Aldous Huxley, *Eyeless in Gaza* (New York: Harper Perennial Modern Classics, 2009), 468. Huxley's statement is like that of Aldo Leopold in *A Sand County Almana* "A thing is right when it tends to preserve the integrity, stability, and beauty of the biotic community. It is wrong when it tends otherwise."

7. In his book *Behave* (p. 326), neuroscientist and primatologist Robert Sapolsky speaks disparagingly of the advent of agriculture: "Which brings us to agriculture. I won't pull any punches—I think that its invention was one of the all-time human blunders…" Sapolsky blames our dependence on domesticated plants and animals for making us vulnerable to droughts, blights, and zoonotic diseases. He blames our settled way of life (our "proximity to feces") for diseases. He blames our food surpluses for class hierarchies. And he thinks agriculture "let loose the dogs of war." He is wrong on all counts. He fails to see that the real cause of all these problems is excessive birth rates. Moreover, just two pages after telling us that hunter-gatherer societies always equitably share their surpluses of meat, Sapolsky tells us that the surpluses generated by agricultural societies lead "almost inevitably" to unequal distribution, "generating socio-economic status differences." Why would that be? After all, it is just as easy to share vegetables as it is to share meat. What gardener doesn't enjoy sharing the fruits of the harvest with friends and neighbors?

8. Jared Diamond, on p. 357 of *The Third Chimpanzee* (New York: Harper Collins Publishers, 1992), states that about 20 percent of

the world's island-dwelling bird species became extinct following the arrival of humans. In North America, the arrival of humans resulted in the extinction of 73 percent of large mammals, while in South America and Australia, the large mammal losses were 80 percent and 86 percent respectively.

9. Malcolm Potts and Roger Short, *Ever since Adam and Eve: The Evolution of Human Sexuality*, p. 237.

10. Jesse Bering, "A Bushel of Facts About the Uniqueness of Human Pubic Hair," *Scientific American*, March 1, 2010. See https://blogs. scientificamerican.com/bering-in-mind/a-bushel-of-facts-about-the-uniqueness-of-human-pubic-hair/.

11. Maurice I. Muoneke and W. Michael Childress, "Hooking Mortality: A Review for Recreational Fisheries," *Reviews in Fisheries Science* 2, no. 2 (1994): 123–56. See also Doug Vincent-Lang, Marianna Alexandersdottir, and Doug McBride, "Mortality of Coho Salmon Caught and Released Using Sport Tackle in the Little Susitna River, Alaska, *Fisheries Research* 15, no. 4 (1993): 339–56.

12. There is an interesting new book by Jonathan Balcombe called *What a Fish Knows: The Inner Lives of Our Underwater Cousins*. It describes the fascinating scientific research on the remarkable cognitive abilities of fish.

13. But what should we do if we encounter a hungry person? Aren't we morally obliged to give her a piece of bread, even though we know this will likely result in her producing still more people destined to be hungry? In other words, shouldn't we "do the right thing now" (feed her) and ignore the "noble goal" of reducing population size? This is a false quandary. The correct answer is that we should give her two things at once: food and an IUD.

14. John Stuart Mill, *Utilitarianism* (1863), chapter 2.

15. This is the first stanza of "Memoir." *The Selected Poetry of Robinson Jeffers* ed. Tim Hunt (Stanford: Stanford University Press, 2001), 518.

16. Henry David Thoreau, *Journal*, March 15, 1860.

17. Henry David Thoreau, *Walden*, "Higher Laws." Despite the strong ethical, environmental, and health arguments in favor of a plant-based diet, it has so far proven difficult to wean people from meat. The anti-meat movement today is about where the anti-tobacco movement was in the 1950s. Back then, medical doctors and famous athletes advertised cigarettes, and millions of children daily inhaled the toxic fumes of their parents' tobacco.

Protection of Pets

Reducing the human population would, in the same proportion, reduce the pet population and its large environmental impacts. In the United States, pet impacts are large because their numbers are large: some 77.5 million dogs and 96.3 million cats.[1] The following table compares the total weight (biomass) of America's dogs, cats, and adult humans.[2]

Species	Average Weight	Population Size	Population Size x Weight (lb.)
American adult	178 lb.[2]	249 million (people over 15)	43,322 million
American dog	35 lb. (est.)	77.5 million	2,712.5 million
American cat	10 lb. (est.)	96.3 million	960.3 million

If we divide the total weight of America's dogs and cats (3,672.8 million lb.) by the total weight of America's adults and teens (44,322 million lb.) we get 0.082 or about 8 percent. So America's dogs and cats equal in weight about 8 percent of the US adult population, which is 20 million adults. This is equivalent to the entire population of Mozambique or Sri Lanka. And since most pets are carnivores, their impacts include all the impacts of the meat industry. Added to this are the impacts of outdoor cats, which kill billions of small mammals and birds every year. The good thing is that pets consume very few *nonfood* resources compared to humans.

Many pets are acquired on a whim or to appease the pleading of children. The purchasers seldom have adequate knowledge of the animal's physical and psychological needs or the amount of time and money that will have to be spent to properly care for it. Children invariably promise to care for the pet but rarely do so unless compelled by parents (which is unpleasant for all). Purchasers of pets rarely stop to consider that the cute puppy or kitten they purchase today will soon grow old and develop infirmities, including incontinence. The result of this ignorance and irresponsibility is suffering for the animals. Many dogs are chained up, beaten, denied adequate shelter and companionship, dumped in kennels when their owners go on vacation, or simply abandoned when their masters find them to be too much trouble and expense. Guinea pigs, rabbits, hamsters, birds, and fish often fare even worse. Even before the animals arrive in the home, they suffer in pet stores and pet breeding facilities and in capture and transport from their wild homes. This abuse is the inevitable result of treating animals as commodities. Pet acquisition is almost always done for selfish motives. The owners (whether adults or children) may be looking for devotion or hoping to win respect by parading a status symbol, or they may want to satisfy a maternal instinct, or feel dominant, or they may want an inexpensive guard for their person and property. In the last case, the dogs may be encouraged to be aggressive, and this results in maulings when they escape confinement.

The maternal instinct often leads women to acquire pets as substitutes or supplements for children. Dog breeders have catered to this appetite by developing "toy" breeds. Rather than continue to exploit animals as baby substitutes, we should form closer-knit communities in which parents of young children will feel comfortable letting women without children help care for them. This would benefit everyone: all adults would have an opportunity to play with children, parents would get some relief, and children would broaden their knowledge and experience. It would also make the abuse of children far less likely, since they would be protected and loved by a whole community instead of being at the mercy of one or two parents who may be incompetent.

If we required would-be pet owners to meet the following requirements, the suffering of pets could largely be eliminated:

- Pass a comprehensive written exam to demonstrate knowledge of how to keep the pet physically and emotionally healthy, and if the pet is a dog or horse, how to train it skillfully and compassionately. Instructional booklets and videos should be made available at public libraries to help people study for the exam.
- Sign a statement acknowledging the average annual and lifetime costs of maintaining the pet and the number of hours of labor (per week, per year, and per average animal life span) that will need to be spent to properly care for the pet. Then sign a promise to provide this care.
- Provide verification of sufficient financial resources to properly care for the animal.
- Purchase an insurance policy to cover the cost of the pet's keep in the event the owner dies or becomes incapacitated, impoverished, imprisoned, deployed overseas, or wearies of the pet and gives it up.

- Have no criminal convictions of cruelty to animals.
- Agree to submit to one unannounced inspection by a pet welfare agent after adopting the pet to ensure that the animal is being properly cared for.

After all these requirements are met, there should be a waiting period of 10 days before people are given control of the pet. This will give them time to reconsider. To further protect pets against cruelty, a generous reward system should be established to encourage tips about abuse, and there should be a well-funded corps of animal abuse investigators.

Pets returned to a humane shelter for whatever reason should be swiftly adopted out. This will be easy once we limit the supply of pets produced by breeding facilities.

All pets should be sterilized. If we decide that pets should continue to be propagated (and I hope we will not), then the breeding of these animals should be restricted to compassionate nonprofit organizations rather than home breeders looking to make a quick profit. Breeds that have congenital health problems should be allowed to die out. Animals produced by breeding organizations should remain under the permanent legal control of the organization, rather than the people who adopt the animals. This will help bring about a conceptual shift from pet ownership as a right to animal guardianship as a privilege.

Adopting the above measures would provide these benefits:

- Eliminate nearly all cruelty to pets, intentional and unintentional.

- Eliminate the problem of abandoned pets (and thus their impacts on wildlife and the economic pressure to euthanize them when impounded).
- Greatly reduce the environmental impacts of the pet food industry, because pets will be far less numerous.

Any community that genuinely wants to protect pets from suffering could enact these measures today. Animal protection organizations, such as the Humane Society of the United States (HSUS) and People for the Ethical Treatment of Animals (PETA), should become strong advocates for these measures.

Those who cannot afford a dog or cat, but who would still like to have contact with these remarkable animals, could volunteer to work at shelters or volunteer to walk other people's dogs. Another alternative to pet ownership would be to improve habitats for wild animals. People can do this in their own backyards, by planting native plants that benefit desirable wildlife. Many communities also have natural area stewardship programs that welcome volunteers. These programs improve habitats for native wildlife. Yet another way people could interact with animals (while simultaneously helping them) is by learning how to rehabilitate injured wildlife.

Finally, if the legal measures proposed above were applied to children as well as pets, we would see a swift end to irresponsible parenthood and child abuse. To perform most jobs in our society, we require formal training, but when it comes to the socially critical and difficult task of childrearing, we require no training at all. That makes no sense.

The reasons for legal intervention in the favor of children, apply not less strongly to the case of those unfortunate slaves and victims of the most brutal part of mankind, the lower animals…What it would be the duty of a human being, possessed off the requisite physical strength, to prevent by force if attempted in his presence, it cannot be less incumbent on society generally to repress.

—John Stuart Mill[3]

Notes

1. United States Humane Society website: www.humanesociety.org.

2. Cynthia L. Ogden, PhD; Cheryl D. Fryar, MSPH.; Margaret D. Carroll, MSPH; and Katherine M. Flegal, PhD, "Mean Body Weight, Height, and Body Mass Index, United States 1960–2002," *Advance Data from Vital and Health Statistics*, no. 347, October 27, 2004. US Centers for Disease Control, Division of Health and Nutrition Examination Survey. The average weight applies to people 20–74 years old.

3. John Stuart Mill, *Principles of Political Economy* (Prometheus Books, 2004), 872–73. Originally published in 1848.

Part 8:

Religion and Overpopulation

What Adam and Eve Did Wrong

The discovery of the causal link between sexual intercourse and birth—two events separated by nine months—was a landmark human achievement. It enabled mankind to begin reproducing with conscious awareness and intent.

In the biblical story of the Fall, Adam and Eve consume a forbidden fruit that opens their eyes to the knowledge that they can create life in their own image—like gods.[1] This knowledge makes them self-conscious about their sexuality, so they cover up their sexual organs.[2] When God observes their newfound shame, he queries them, and they admit to eating the forbidden fruit (though Adam blames Eve, and Eve blames the serpent). God decides that their punishment will be to suffer an eventual death. (Birth and death must balance.) So God expels Adam and Eve from the Garden, depriving them of access to a fruit that confers immortality. As further punishment, God declares that Eve will have to bear her children in pain, and Adam will have to support them by the sweat of his brow.

Adam and Eve were not alone. They had neighbors. When God informed Cain that he would be expelled from Eden for killing Abel, Cain protested that the people out there would kill him. God responded by giving him a special mark to protect him from harm. Cain then moved out, found a wife, and fathered children.[3]

Adam and Eve did nothing wrong by discovering that they could make life in their own image. Knowledge by itself is never harmful. Nor was it wrong for them to produce children. Reproduction, after all, was part of God's plan for all creatures (which, of course, is why all creatures have reproductive organs and sexual desire). Adam and Eve's error was simply this: they turned their backs on their neighbors and created their own neighbors. They thought it would be much easier to love people created out of themselves.

In revealing what went wrong, the story of Adam and Eve shows us the way out. Instead of turning inward, we have to turn outward, back toward our neighbors. We have to love our neighbors as ourselves and our children as our neighbors. It is only when we love our children as our neighbors that the abuse of children will end. Who would dare treat a neighbor the way so many parents treat their children?

It is ego (constructed of memory) that aspires to be immortal. Those who imagine they can survive death by producing children and molding them into their own image have confused their egos with their souls. True immortality—the Tree of Life—is realized when our egos fade away, allowing us to perceive our integral connection with the whole world.

If we ever do manage to love our neighbors as ourselves and our children as our neighbors, sexual shame will vanish:

> His disciples said, "When will you become revealed
> to us and when shall we see you?" Jesus said, "When
> you disrobe without being ashamed and take up
> your garments and place them under your feet like

little children and tread on them, then [will you see] the son of the living one and you will not be afraid."
—Gospel of Thomas (37)[4]

Notes

1. Because the forbidden fruit radically expanded knowledge, medieval artists sometimes represented the Tree of Knowledge as an *Amanita muscaria* mushroom. This mushroom contains a powerful mind-expanding chemical called muscimol.

2. Adam and Eve's first clothing consisted of fig leaves stitched together. The association of figs with sexuality and fertility is an ancient one. It stems from the peculiar womb-like fruit of figs. This fruit (called a syconium) is structurally like a large mulberry turned inside out. (Figs and mulberries belong to the same botanical family.) Fig trees bear their male and female flowers on separate trees, and they are pollinated by a tiny wasp *(Blastophaga psenes)*. Once this wasp enters the young syconium, it pollinates some of the flowers and lays eggs in the rest.

3. Adam and Eve's neighbors are illustrated in an outstanding condensed version of the first book of the Bible written and illustrated by R. Crum: *The Book of Genesis Illustrated* (New York: W. W. Norton & Co., 2009).

4. *The Gospel of Thomas*, trans. Thomas O. Lambdin, in *The Nag Hamadi Library in English* (New York: HarperCollins, 1992), 130.

Humanae vitae Examined

The Roman Catholic Church is the world's largest religious organization, with more than 1 billion members. Its hierarchy has long been the world's foremost opponent of contraception and thus a major obstacle to halting and reversing population growth. It is therefore important to examine the Church's teaching on contraception.

Humanae vitae ("of human life") is the official Roman Catholic document that defends the Church's centuries-old opposition to all methods of birth control except sexual abstinence and the unreliable rhythm method. *Humanae vitae* was introduced by Pope Paul VI in July 1968. All subsequent popes have endorsed it.

Humanae vitae teaches that reproduction is a three-way effort involving a man, a woman, and God. The duty of the man and woman is to let their sperm and egg cells unite freely (or at least not to use any "artificial" methods to impede this union).[1] God's role is to implant a soul in each zygote at the moment of conception.

The notion that an omnipotent God requires human cooperation to effect fertilization is, of course, absurd. If God's will can be thwarted by a condom, then God is impotent and not a god. Another problem: if it is wrong to prevent God from inserting a soul into a fertilized egg, then it should also be wrong to prevent God from retrieving a soul whenever he wishes. In other words, if Catholic dogma were logically consistent, good Catholics would have to refrain from all medicines and surgeries—anything that would "artificially" impede God's effort to reclaim their souls. And if they did this, their death rates would soar, bringing them into alignment with birth rates, thereby solving the problem of population growth the natural way—God's way. By embracing only the creative side of God, and conveniently ignoring the destructive side, *Humanae vitae* shows itself to be unbalanced and distorted.

Not only is it theologically inconsistent for the Church to oppose birth control while supporting death control, but the overpopulation that results from the excess of births over deaths greatly increases poverty, disease, violence, and environmental degradation. These are all evils that the Church claims to oppose. *Humanae vitae* therefore pits the Church against itself. It prevents the Church from acknowledging the obvious: that fewer children per family would mean greater material wealth for each family member, along with less disease, less violence, and less environmental harm.

The Church's contention that God wants women to be open to pregnancy whenever they engage in sex defies logic. If that had been God's intention, he would have restricted human interest in sex to the brief period of fertility (6 days per month). The fact that he made humans keenly interested in sex throughout the month and throughout their lives can only mean one thing: that he wanted and expected humans to engage in sex frequently *regardless of reproduction*.

Although most Catholics, to their credit, have ignored *Humanae vitae*,[2] the political power of the hierarchy has led the governments of some Catholic nations to restrict access to contraception.[3] Wherever this has occurred, higher abortion rates have inevitably followed. Because these abortions are illegal, they imperil the mother's health and frequently cause her death.[4] Thus the Church's opposition to contraception, while claiming to be based on respect for life, has served to increase the number of aborted fetuses, dead mothers, and orphaned children.

Even if the Catholic bishops recognized that *Humanae vitae* is theologically and morally wrong, they would still be reluctant to publicly admit the Church's error. They would worry that such an admission might unsettle the faithful, leading them to question other doctrines, thereby undermining the hierarchy's authority and power.[5] But a hierarchy that is preoccupied with its own authority and power is a hierarchy full of pride, and pride (says the Church) is a deadly sin.

In sum, *Humanae vitae* is theologically untenable and morally indefensible. Its effect has been to promote poverty, suffering, premature death, and the severe degradation of the natural world (God's creation). By the Church's own moral standards, the teaching of *Humanae vitae* should be condemned as a mortal sin.

> Reason and free enquiry are the only effectual agents against error. Give a loose to them, they will support the true religion, by bringing every false one to their tribunal, to the test of their investigation. They are the natural enemies of error, and of error only.
> —Thomas Jefferson[6]

Notes

1. Jared Diamond commented that the rhythm method would work very well for most mammalian species, because these species conspicuously signal their brief period of ovulation. However, the rhythm method is singularly inappropriate for humans since human

ovulation is completely concealed. See *The Third Chimpanzee* (New York: HarperCollins Publishers, 1992), 78.

2. Among sexually active Catholic women who are not pregnant, postpartum, or trying to get pregnant, 2 percent use Church-approved natural family planning (the rhythm method), while 68 percent use effective contraceptive methods not approved by the Church. Of the latter group, 32 percent use sterilization (24 percent female and 8 percent male), 31 percent use oral contraceptives, and 5 percent IUDs. Of the remaining 30 percent, about 11 percent use no contraception at all, and the rest use non-fail-safe contraception, such as condoms and coitus interruptus. US Catholic women use contraception in nearly identical proportions to women of other mainstream religions. See "Countering Conventional Wisdom: New Evidence on Religion and Contraceptive Use," by Rachel K. Jones and Joerg Dreweke (New York: Guttmacher Institute, 2011). This article can be viewed at https://www.guttmacher.org/report/countering-conventional-wisdom-new-evidence-religion-and-contraceptive-use.

3. The Philippines is an example of a country that has bowed to pressure from the Catholic Church. The Philippine legislature in 2016 removed all government funding for contraception. The population of the Philippines is growing at a dismaying rate: in 1970 (around the time high-yield rice was introduced), the population was 36.6 million, but by 2016 it was 103 million (2.8 times larger). Unchecked breeding has caused Filipinos to ruin their environment and trap themselves in permanent poverty. The Catholic Church bears a large share of responsibility for this moral and environmental catastrophe.

4. The largest maternity hospital in Bogotá, Colombia, uses half its beds for women suffering the complications of illegal abortion. See F. Jaffe, B. Lindheim, and P. Lee, *Abortion Politics: Private Morality and Public Policy* (New York: McGraw-Hill, 1981). In Nicaragua in the 1980s, 45 percent of the beds in a large maternity hospital in Managua were occupied by women suffering from illegal abortions.

Illegal abortions are the number one cause of maternal death in Nicaragua. See http://countrystudies.us/nicaragua/29.htm.

5. Garrett Hardin, in an enlightening essay called "The Ghost of Authority" (in *Naked Emperors: Essays of a Taboo-Stalker*), points out that authority and conscience are the same thing. Adults *decide* to obey authority because their *conscience* tells them to. There is no authority above conscience.

6. Thomas Jefferson, *Notes on the State of Virginia (1787)*, query XVIII.

Why the World's Major Religions Have Ignored Overpopulation

All religions concern themselves with ethics and human welfare, yet no religion has ever acknowledged the central contribution of overpopulation to human misery and vice. Religions have instead blamed human suffering on original sin, or the Devil, or bad karma, or divine inscrutability. One reason religions have ignored overpopulation is obvious: they want more members. Another reason is historical: when these religions were formed, overpopulation was held in check by wars, epidemics, famines, and infanticide. Those checks are no longer effective, yet religions continue to ignore overpopulation. Part of the answer lies in their creeds. In this chapter, I'll briefly examine the creeds of the world's four largest religions to shed light on why they have ignored overpopulation.

Christianity, the world's largest religion, began as an apocalyptic Jewish sect whose members believed that "within a generation" a cosmic judge would descend from the clouds, accompanied by angels (Matthew 25:31). This judge would cast the wicked into hell, resurrect the righteous dead, and establish a divine kingdom here on earth with Jerusalem as its capital. But those predictions never came to pass. Instead, the principal leaders of the sect—John the Baptist, Jesus, Peter, James, and Paul—were, one by one, arrested and executed.[1] Then, in the year 70, the Roman army sacked and burned Jerusalem and demolished the Temple (God's house). These misfortunes must have perplexed and disheartened the members of the sect. But they managed to preserve their faith, because they came to believe that they would go to the Kingdom when they died. "Thy Kingdom come" became "I'll come to thy Kingdom."[2] This belief shift proved fortuitous, because it enabled Christians to begin winning many converts (including non-Jews) by promising them eternal life after death.

The Christian focus on gaining admittance to an invisible posthumous kingdom meant that Christians had no long-term commitment to this world (despite calling it God's creation). From a Christian perspective, this world is a stopover on the way to heaven or hell—a kind of moral proving ground filled with temptations and wickedness. Given this deeply negative attitude toward the world, it is not surprising that Christianity has never called upon its members to exercise reproductive restraint. In fact, the largest Christian denomination—Roman Catholicism—actively promotes overpopulation by opposing contraception and abortion.

Despite Christianity's otherworldly focus, there have been occasional thoughtful Christians who did call for reproductive moderation. One of them was Thomas Malthus, an Anglican cleric. He made the following appeal to his fellow Christians:

> To the Christian I would say that the scriptures most clearly and precisely point it out to us as our duty, to restrain our passions within the bounds of reason; and it is a palpable disobedience of this law to indulge our desires in such manner, as reason tells us, will unavoidably end in misery.[3]

In our own time, economist Herman Daly and theologian John Cobb have repeated the Christian call for restraint:

> Our ability and inclination to enrich the present at the expense of the future, and of other species, is as real and as sinful as our tendency to further enrich the wealthy at the expense of the poor. To hand back to God the gift of Creation in a degraded state capable of supporting less life, less abundantly, and for a shorter future, is surely a sin.[4]

Islam, the world's second largest religion, has its roots in both Judaism and early Christianity. Muslims believe that the dead will remain in their graves until the Day of Judgment, whereupon they will be sent to heaven or hell according to their behavior during life. There are two exceptions to this: Muslims who die fighting for Islam go straight to paradise, and the enemies of Islam go straight to hell. Because Muslims don't claim to know when the Day of Judgment will occur, one might think they would take better care of the world in the interim than Christians have. But the sad reality is that all Muslim nations are severely overpopulated and environmentally degraded. Due to overpopulation, millions of Muslims now seek to escape to more prosperous Christian lands (and many succeed). Those left behind face a nearly hopeless future. High levels of unemployment mean that large numbers of young Muslims will never have the means to marry. Frustrated and angry, it is no surprise that some of them are drawn to violent apocalyptic cults that promise unlimited sex in paradise to those who die fighting infidels (as well as plenty of earthly sex in the interim—sometimes with captured slaves). Having

failed to control their birth rates, Muslims now face decades of political turmoil, war, poverty, suicide bombings, and mass exodus to
non-Muslim lands (where they continue to breed excessively).[5]

Until recently, there was one bright spot in the Muslim world:
the nation of Iran. From 1990 to 2012, Iran operated a birth control program. This program lowered Iran's growth rate to 0.7 percent (which is the same as the current US growth rate). However,
in 2012, Supreme Leader Ali Khamenei declared that Iran's family
planning program was "wrong." Following his decision, funding for
birth control programs was slashed. It remains to be seen how much
this will boost Iran's birth rate.

Hinduism, the world's third largest religion, is a colorful mix of
many Indian traditions, rituals, and gods. Hindus believe that when
they die they will be reborn here on earth (rather than spending
eternity in an extraterrestrial heaven or hell, or lying in their graves
until the Day of Judgment). Despite their belief in a continually renewed life on earth, Hindus have never shown much regard for the
sustainability of earth's ecosystems—as is evident from India's huge
and explosively expanding population. Rather than urge people to
work together to rationally solve problems, Hinduism encourages
the private pursuit of salvation through the renunciation of adult
responsibilities and the adoption of an ascetic lifestyle. Hinduism
also teaches that the suffering of the poor is a deserved consequence
for wrongs committed during previous lives. This notion of karma
conveniently serves the interests of the ruling elite by encouraging
the poor to blame themselves (or rather their former selves) for their
current woes. How much better it would be if the poor were taught
the truth: that their suffering stems from having produced more

children than they (and their environment) can adequately support, as well as their exploitation by the higher castes.

Buddhism, the world's fourth largest religion, is an interesting case. The following table compares the teachings of Buddhism with the understandings of biology:

The Four Noble Truths of Buddhism	The Four Noble Understandings of Biology
Life entails suffering.	Overpopulation entails suffering.
Suffering is caused by desire.	Overpopulation is caused by the instinctive desire to breed, which eventually leads to the harsh pruning process of natural selection.
Suffering is not inevitable: desire can be overcome.	For humans, nature's painful pruning process is not inevitable. Unlike other animals, we have the capacity to use reason and contraception to outmaneuver both the sex drive and the maternal drive.
The way out of suffering is the eightfold path: right view, right attitude, right speech, right action, right livelihood, right effort, right awareness, right concentration.	The way out of suffering is to adopt an effective method to regulate birth rates and immigration so as to attain (and then maintain) an optimal population size. Such a size would allow humans to enjoy a good life without causing undue harm to other species.

If we interpret "right view" to mean "understanding the dangers of overpopulation" and "right action" to mean "adopting an effective system to regulate birth rates and immigration," then Buddhism can be reconciled with biology. However, neither the Buddha nor his followers ever made these critical connections (although the Buddha did wisely establish a policy of celibacy for monks and nuns). To their credit, Buddhists have lower birth rates than Muslims, Hindus, and Christians, but all Buddhist nations are still overpopulated. The pressures of overpopulation are now giving rise to Buddhist nationalism and the persecution of non-Buddhists in Sri Lanka, Burma, and Thailand.[6]

Religions have been urging people to lead ethical lives for thousands of years, with little success. People are still selfish and cruel, and the worst of the lot are often the most religious. There are many reasons why good ethics have never taken hold, but overpopulation is the main one. As long as people must struggle to survive, they will never behave well.

This is not to say that religion has nothing to offer. Many people find comfort in the notion that there are invisible powers who care about them and will come to their aid if they pray or perform the right rituals. Religion can also reinforce good ethics and provide people with a supportive community and opportunities to help others. For those with obsessive-compulsive tendencies, the repetition of religious rituals can help to allay anxiety.[7] But religious beliefs and rituals can never transform people at a deep level. Only direct spiritual experience can do that. Moreover, it is but a small step from believing religious nonsense to believing political nonsense. In America, we have recently seen millions of evangelicals flock to Donald Trump (a man without religious beliefs), blindly embracing his lie that the 2020 US presidential election was stolen. That poses a threat to democracy.

The most effective way to reduce the power of religion would be to establish governments that provide for the basic needs of their people. When people have adequate income, health care, and security, they no longer have any incentive to beg invisible gods. In Europe, where most governments now provide a social safety net, religious attendance has fallen sharply, and with it the ability of religious leaders to influence government policy.

Although no religion has ever called upon its followers to regulate reproduction, religions have often unintentionally helped to reduce population growth. One way they have done this is by fostering wars. The Muslim conquests of the 7th century and the Crusades of the 12th and 13th centuries are well-known examples. In fact, throughout history, religious leaders have been eager supporters of their nation's wars. Currently, for example, Russian Patriarch Kirill is voicing strong support for Putin's vicious war against Ukrainians. And older Americans may recall that two cardinals of the Catholic Church—Spellman of New York and McIntyre of Los Angeles— were outspoken supporters of the Vietnam War. This was a war that prematurely subtracted at least 2 million people from the world's population (including nearly 60,000 young Americans). Spellman and McIntyre thought it was a splendid idea to send young Americans halfway around the world to kill Asian peasants who posed no threat to them. Yet these same cardinals condemned birth control as a sin. Think about that. They thought it was a sin to prevent the nonexistent from existing, but commendable to violently force the existing into nonexistence! Thankfully, not all religious leaders of that era were so benighted. Some spoke out forcefully against the Vietnam

War, including Bishop James Pike, Rev. William Sloane Coffin Jr., Rev. Martin Luther King Jr., and Bishop Thomas Gumbleton.

Another way religions have unintentionally reduced overpopulation is by sequestering large numbers of men and women in monasteries and convents. This sequestration not only prevents them from breeding, but frees up resources for the rest of society, because the monks and nuns generally live quite simply. In Thailand, Burma, and Sri Lanka, where large numbers of young men spend a period of time in a monastery, the effect is to postpone marriage, which helps significantly to lower the birth rate. Mormons accomplish the same thing by sending their young men and women away for two years of missionary work (although they tend to make up for it when they come home). In Tibet, prior to the Chinese takeover, about 20 percent of men were celibate monks, and many Tibetan women were nuns. The Tibetan birth rate was further lowered by the widespread practice of fraternal polyandry, in which two or more brothers married the same woman. Polyandry lowers the birth rate by reducing the number of eligible men available for other women to marry (assuming a balanced sex ratio).[8]

Some American Indian tribes also had a way to restrict population size (besides war). Their ceremonial custom of having men sweat in a sweat lodge reduced male fertility, because high temperatures kill sperm. Scandinavians and Russians no doubt obtain the same benefit from their saunas.

Notes

1. The arrest of Jesus was prompted by an altercation that took place in the temple compound—an event related by all four gospel authors. During this scuffle, Jesus liberated some animals intended

for sacrifice and overturned the tables of the money changers. The role of the money changers was to exchange Roman coins (which bore the "graven image" of Caesar) for the Temple's own coins. This coin exchange was considered necessary because Caesar had been declared a god ("son of the divine Julius"), and it would have been sacrilegious for Jews to use such coins in the Temple. The function of the temple coins was to purchase animals for sacrifice. These sacrifices were performed by a priest (kohen), who would slit the animal's throat. Once the victim bled to death, the priest would burn some of its fat on the altar as a token offering to God. After that, he would take a cut of meat for himself, and return the rest to the owner. In this way, the faithful obtained kosher meat, along with reassurance that God would be more inclined to fulfill their prayers thanks to the burnt offering. But animal sacrifice also meant that Judaism's most sacred sanctuary was a slaughterhouse—a reeking hellhole of terror, cruelty, and death. Animal sacrifice also implied that God was a carnivore who needed to be fed like a tiger in a zoo. Not surprisingly, more than a few Jews found this pseudo-religious butchery disgusting and rejected its demeaning implications about God. Among the scriptural passages that reflect this opposition are the 50th Psalm, Hosea 6:6, 8:13, 14:2, and Micah 6:6–8. The members of Jesus's sect were among those who opposed animal sacrifice, and there is considerable biblical and nonbiblical evidence that they were vegetarians. In opposition to Jesus's group stood the Sadducees, who controlled the lucrative temple meat trade. The Sadducees clearly saw that opposition to animal sacrifice posed a threat to their profits, so they told the Roman authorities that Jesus was a subversive. The Roman governor, Pontius Pilate, promptly had him crucified. Years later, a schism developed among Jesus's followers when Paul (who had a small following in Syria that included some non-Jews) declared that there was nothing wrong with eating meat. This shocked the core group in Jerusalem, led by James, Peter, and John (known as "the three pillars"). James (the brother of Jesus) sent a delegation to Paul to try to rein him in. Biblical scholars call this confrontation the "incident at Antioch." It is described in Galatians 2:11–14. Paul's surviving letters provide fascinating glimpses into this conflict. See Romans 14:2, Romans 14:14, Romans 14:20–21, 1 Cor 8:4, 1 Cor 8:7, 1 Cor 8:13, 1 Cor 9:3, 1 Cor 10:25, and 1 Cor 10:27. See also Acts 10:10–16. In the end, Paul's faction triumphed, and Christianity became a religion of meat eaters (a fact that helped to ensure the religion's widespread success). Instead of sacrificing animals in their churches, Pauline Christians began the custom of ritually eating the sacrificed body of Jesus and drinking his blood. (The latter practice is remarkable given the religion's origin in Judaism, which strictly forbids blood consumption, not to mention cannibalism.) In contrast to the Pauline Christians, the original Jewish followers of Jesus (who were never called Christians, but Nazoreans) continued to oppose animal sacrifice and remained faithful to a vegetarian diet. The Nazoreans eventually became known as the Ebionites, and under that name they persisted for a few centuries. Interestingly, in the Semitic languages, Christians are still called Nazoreans: *notzrim* in Hebrew and *nasrani* in Aramaic and Arabic.

2. The switch from "thy Kingdom come" to "I'll come to thy Kingdom" was as if Didi and Gogo finally wearied of waiting for Godot and decided to pay him a call—but first they had to die!

   The assertion that the Kingdom would come "within a generation" was a testable hypothesis. Either it would arrive within a generation, or it would not. When it did not arrive, the claim was refuted, and the believers looked foolish. But once the early Christians adopted the belief that they would go to the Kingdom when they died, they had a belief that was untestable and thus irrefutable. That made it very appealing. In the historical competition among the world's religions, the belief in a postdeath heaven or hell has proven to be the fittest dogma. Second place goes to reincarnation, and third place to apocalyptic religions that predict the arrival of a messiah who will rule the world. The reason apocalyptic cults continue to have appeal (especially among the poorest members of society), is that they do not require their members to die before experiencing the delights of the Kingdom. They need only be patient for a while (and keep contributing to the church).

3. Thomas Robert Malthus, *An Essay on the Principle of Population* (1798), book 4, chapter 2.

4. Herman Daly, *Beyond Growth: The Economics of Sustainable Development* (Boston: Beacon Press, 1996), 222.

5. In North America, the total fertility rate for Muslims was 2.7 in the period 2010–2015, in contrast to the regional average of 2.0 (http://www.pewforum. org/2015/04/02/north-america/). Throughout the world, Muslims are out-reproducing every other religious and nonreligious group. People interested in learning more about Muslim fecundity should read "Why Muslims Are the World's Fastest-Growing Religious Group" by Michael Lipka and Conrad Hackett. Their article was published by the Pew Research Center on April 26, 2015, and is available at https://www.pewresearch. org/fact-tank/2015/04/23/why-muslims-are-the-worlds-fastest-growing-religious-group/. Here are some excerpts from this article:

"While the world's population is projected to grow 35% in the coming decades, the number of Muslims is expected to increase by 73%—from 1.6 billion in 2010 to 2.8 billion in 2050. In 2010, Muslims made up 23.2% of the global population. Four decades later, they are expected to make 29.7% of the world's people.

By 2050, Muslims will be nearly as numerous as Christians, who are projected to remain the world's largest religious group at 31.4% of the global population.

The main reasons for Islam's growth ultimately involve simple demographics. To begin with, Muslims have more children than members of the seven other major religious groups analyzed in the study. Each Muslim woman has an average of 3.1 children, significantly above the next-highest group (Christians at 2.7) and the average of all non-Muslims (2.3). In all major regions where there is a sizable Muslim population, Muslim fertility exceeds non-Muslim fertility.

The growth of the Muslim population also is helped by the fact that Muslims have the youngest median age (23 in 2010) of all major religious groups, seven years younger than the median age of non-Muslims (30). A larger share of Muslims will soon be at the point in their lives when people begin having children. This, combined with high fertility rates, will accelerate Muslim population growth."

6. In Burma (Myanmar), violence has broken out between the native Buddhist population and the Muslim Rohingya minority who immigrated to Burma from neighboring Bangladesh. In May 2015, the Burmese government passed a family planning law that mandated a three-year waiting period between births. Some in the West condemned this legislation because they believed it was targeted at the Rohingya, who, like most Muslims, have a high birth rate. Regardless of whether or not this is true, the legislation lacked any enforcement provisions and was therefore little more than a recommendation. But that didn't stop Human Rights Watch and the UN rapporteur to Burma from reflexively condemning it. Subsequently, we know, the Burmese army terrorized and murdered many Rohingya, driving most of the survivors into neighboring Bangladesh.

7. For a discussion of the connection between obsessive-compulsive disorder (OCD) and religious rituals see the essay "Circling the Blanket for God" in Robert Sapolsky's book *The Trouble with Testosterone* (New York: Touchstone, 1997).

8. Melvyn C. Goldstein, "New Perspectives on Tibetan Fertility and Population Decline," *American Ethnologist* 8, no. 4: (November 1981), 721–38.

9. Malcolm Potts and Thomas J. Hayden, *Sex and War* (Dallas: Ben Bella Books, 2008), 360.

Part 9:

Wrapping It Up

No Despair

The future is never hopeless: solutions are always out there. In December 2008, a protest took place in Salt Lake City against the sale of oil and gas leases on public lands in Utah. The mood of the protesters was grim, for it seemed nothing could be done to stop the leases. One of the protesters, a young man named Tim DeChristopher, decided to enter the building where the auction was about to be held. At the registration desk, he was asked if he was there to bid. On an impulse, he said "yes," and became Bidder 70. As the bidding began, he thought he might stand up and make a speech, but he sat quietly until he saw a friend from his church in tears over the sale. That moved him to join the bidding. At first, he merely pushed up the price of the parcels, but when almost half the parcels had been sold, he decided to start winning them. He won all subsequent parcels, until it became apparent that he was an imposter. The auction was then canceled. The incoming Obama administration took office before a new auction could be scheduled. Upon review of the parcels in question, Secretary of the Interior Ken Salazar dismissed the auction, declaring that the Bureau of Land Management had cut corners and broken many of its own rules, including a crucial statute requiring all federal agencies to evaluate the impacts on climate prior to auctioning off public lands for the purpose of energy development. In return for his inspired and heroic defense of the land, Tim DeChristopher was prosecuted and sentenced to two years in prison.[1]

Hopelessness occurs when our neural circuitry becomes clogged. This mental sluggishness prevents us from seeing the many options and opportunities that are always out there. If the clogging goes unchecked, the options we perceive may eventually shrink to only one: suicide. There are lots of ways to get the juices flowing again, ranging from reading *Don Quixote* to taking a psychedelic.[2] Exercise and diet are also important, as are friends. When bad things happen, instead of bemoaning our fate we can choose to view our woes with curiosity, being open to whatever opportunities they may present. There is neither complacency nor submission in this, just a desire for insight.

Environmentalists could help allay despair—their own and society's—by developing a clear and persuasive vision of a good society. Without such a vision, the best they can do is issue gloomy forecasts that the public defensively tunes out. Environmentalists need to be able to show how real democracy can be created, how the wealth generated by machines can be equitably distributed, how small family farms can be restored, how the mass media can do a better job of uncovering and conveying truth, how people can develop a love of learning and the habit of thinking critically, how people can be persuaded to adopt a more healthful diet, how health care can be provided efficiently for all, and how the divisive human impulse to identify with groups can be supplanted by the spiritual recognition that we are all one beneath the surface. The environmental movement today is not even attempting to meet these challenges. Instead, it is a shallow movement, focused entirely on mitigating the *symptoms* of overpopulation, the *symptoms* of animal exploitation, and the *symptoms* of consumerism. Until environmentalists are finally willing to admit that overpopulation, animal exploitation, and consumerism are the prime drivers of environmental ills, they cannot heal the world:

> …the people and organizations we look to for guidance and leadership about what we need to do to heal the Earth and live sustainably are not telling us the truth. We get part of the story, an incomplete understanding, and a plan that will not succeed.
>
> —Will Anderson[3]

Fortunately, the public's attitudes can sometimes change rapidly. Back in 2013, when the first edition of this book was published, few people imagined that within a few years, several American states would legalize marijuana, and the US Supreme Court would declare gay marriage legal throughout the land, and a major candidate for the US presidency (Andrew Yang) would call for a universal basic income. We are now hearing calls to end America's extreme disparities in wealth, to end mass incarceration, to put a halt to mass immigration, to end gerrymandering, to reform our corrupt system of campaign finance, and to reform trade agreements that have ruined American industries and increased unemployment. So there is reason to hope that the public's attitudes toward overpopulation, animal exploitation, and consumerism will undergo a similar rapid transformation. There are already hopeful signs: A solid majority of the US population opposes mass immigration, and roughly 5 percent of Americans have stopped eating animals.[4] Whenever we feel inclined to slip into cynicism and despair, we should remember that hopelessness is a self-fulfilling trap. In *Limits to Growth: The 30-Year Update*, the authors list five tools we can use to heal the world:

> They are: visioning, net-working, truth-telling, learning, and loving. It seems like a feeble list given the enormity of the changes required. But each of these exists within a web of positive loops. Thus their persistent and consistent application initially by a relatively small group of people would have the potential to produce enormous change—even to challenge the present system, perhaps helping to produce a revolution.[5]

Notes

1. This account of Tim DeChristopher's defense of the land is from http://www.peacefuluprising.org/tim-dechristopher/tims-story. Tim's action was a kind of judo ("gentle way"). He used the enemy's strength to defeat him.

2. Albert Hofmann, the Swiss chemist who discovered LSD and first isolated psilocybin from Mexican sacred mushrooms, believed that an extremely low dose of LSD (one below the threshold of awareness) would make an ideal antidepressant. Although government-imposed impediments to psychedelic research have so far prevented this idea from being tested, in December 2016 the first study demonstrating the ability of psilocybin to alleviate depression was published. See http://www.nytimes.com/2016/12/01/health/hallucinogenic-mushrooms-psilocybin-cancer-anxiety-depression.html?_r=0. Ayalet Waldman's book, *A Really Good Day*, recounts her success in using microdoses of LSD to overcome depression. It was reviewed in *The Atlantic* https://www.theatlantic.com/health/archive/2017/01/ayelet-lsd-microdosing/513035/.

3. Will Anderson, *There is Hope: Green Vegans and the New Human Ecology* (Hants, UK: Earth Books, 2015), 176.

4. Researchers have found that the most accurate way to determine the percentage of vegetarians in a population is to ask people what they ate yesterday. This is because many people who claim to be vegetarians actually eat fish or poultry.

5. Donella Meadows, Dennis Meadows, and Jørgen Randers, *Limits To Growth: The 30-Year Update* (White River Junction, VT: Chelsea Green Publishing), 271. For a humorous take on the difficulty of getting people to save themselves even when shown a way out, see Doug Stanhope's brilliant 2012 standup performance, "Before Turning the Gun on Himself." It can be found online.

Prospects for Implementation

Economist Kenneth Boulding pointed out that ideas and plans are like genotypes (DNA), whereas the physical and social structures we build from these ideas and plans are the phenotypes upon which natural selection acts.

A population-stabilization plan like the one presented in this book could be converted to reality by any organization that regulates its membership, regardless of size. This could be a religious organization, a co-op, a tribe, or a nation. The members of the group would limit themselves to two children per couple (unless couples acquired additional credits by donation or purchase). If the organization desired to increase its membership, it would do so by recruitment alone instead of reproduction. If a credit-based plan were adopted by a nation, the plan could be applied to the whole nation or restricted to a region of special environmental sensitivity. If restricted to an environmentally sensitive area (such as region of exceptionally high biodiversity), only people who agreed to abide by the conditions of a two-credit system would be allowed to live there.

The value of regulating population size even in small communities is illustrated by the history of The Farm, an intentional community founded in Tennessee in the early 1970s by 250 idealistic hippies. The founders dreamed of establishing a loving, self-sustaining, environmentally sound community that could serve as a model for the rest of society. But that dream was thwarted by demographic naivete. The Farm had an "open-gate" immigration policy: practically everyone who arrived was permitted to stay, provided they were willing to work and not be disruptive. It was also Farm policy to encourage marriage and childbearing. They even invited pregnant women from around the country to give birth at The Farm with the option of leaving their baby behind. More than a few women took them up on this generous offer. The combined effect of these policies was to rapidly swell the population to 1,500, which was unsustainable. Residents lived in poverty and lacked adequate medical and dental care. Eventually the more productive members grew weary of supporting the less productive. They also recognized that it was unfair to impose poverty on their own children. These attitudes finally led to the "changeover": the replacement of communal property with private property. This transformation was followed by a large exodus (emigration). Today, The Farm has a much smaller population, and supports several environment-friendly businesses and charities.[1]

In the United States, a credit-based population-stabilization plan would be opposed by some groups and supported by others. Opponents would include business organizations whose profits are linked to population growth, as well as conservative members of the Christian, Jewish, and Muslim faiths, especially the Roman Catholic hierarchy. Some libertarians might also object to a population-stabilization plan, because they don't like government interference in personal affairs (a view I agree with, except I don't consider reproduction to be a purely private affair once the number of children exceeds two). On the other hand, libertarians love free markets and generally disapprove of free rides, so they would be drawn to those aspects of the plan. The plan would also be supported by all who would likc to profit from selling one or both of their population credits. Demographic data show that as of 2014, 15 percent of American women aged 40 to 44 have never produced a child,[2] and 18 percent have produced only one child.[3] Add them together, and 33 percent of American women could profit handsomely from selling their unused credits. The percentage of men who would profit would likely be about the same. Most of these people would support the two-credit plan out of economic self-interest if nothing else.

Another group that would likely support the two-credit plan would be people over 50 who have produced at most one child. Although these people, being over 50, could not profit from selling a credit or two, they would nonetheless support the two-credit system, because they know they would have profited if they were younger.

Two other groups that would be attracted to the plan are environmentalists,[4] and the 3.5 percent of the US population who are illegal immigrants. The latter would support the plan because it would grant them immediate US citizenship (albeit without giving them population credits). Although illegal immigrants cannot vote, their legal relatives can. And these legal relatives are plentiful. About 9 percent of the US population is foreign born *and* legal. (The foreign-born total is nearly 15 percent.) Moreover, even illegal immigrants can influence legislators through letters.

Many women would also be drawn to the plan, because of its woman-friendly provisions, such as full control over reproduction, easy access to the means of controlling it, and the near elimination of rape by strangers.

Additional supporters would be all the people who are simply fed up with the problems of population growth, which include traffic congestion, expensive housing, polluted air and water, crowded recreation areas, urban sprawl, and so on. If we add all these people together, it is quite plausible that more than half of the US population would support the two-credit system. But enacting the plan would require that Congress represent the will of the people rather than the will of the financial sector and the corporations, and it would require that the mass media honestly present the plan to the people instead of using deception and scare tactics to turn people against it. Unfortunately, America's mass media are controlled by the same powerful business interests that control the government—interests that favor unsustainable population growth. But the internet now provides a way to get factual information directly to the people, bypassing the corporate media.

China would be an excellent candidate for a credit-based population-stabilization plan. The Chinese people are well informed about the problems of overpopulation and experience them every day. A credit-based population-stabilization plan would have substantial advantages over China's current three-child policy (which replaced the two-child policy in 2021):

- The plan would allow the government to make precise adjustments to population size simply by retiring or recycling more credits. This would be far more efficient than using a system of incentives and disincentives to try to change breeding practices.
- Having a highly accurate vital statistics database would permit better planning.
- There would be no more forced abortions, an ugly practice that has occurred in some districts and tarnished China's international reputation.

- Having a DNA database of the entire Chinese population would provide a powerful tool for medical research.

Island nations that have educated populations and are prosperous enough to afford a national DNA lab, a national vital statistics database, and a network of registrars would be excellent candidates for the two-credit plan. Examples would be Japan, New Zealand, Australia, Taiwan, Ireland, and Iceland. The geographic isolation of islands makes it easy for them to restrict immigration and ensure national defense, even with a shrinking population.

A population-stabilization plan need not be perfect before it is implemented. Any plan that offers a clear improvement over the status quo should be undertaken. Plans can always be adjusted once in operation. A credit-based system is highly flexible, lending itself to whatever modifications people choose to make.

Adopting the two-credit plan would not mean abandoning any of the existing approaches to slowing population growth. On the contrary, we should redouble efforts to educate women in poor countries to have smaller families (as Population Media has been doing successfully with TV serial dramas), develop better long-term contraceptives (as the Gates Foundation has proposed to do), make contraceptives more readily available at lower prices (as Fòs Feminista—formerly International Planned Parenthood Federation—does), defend access to safe abortion (as NARAL strives to do), and we should work to eliminate mass immigration (as various organizations are trying to do). Although all these approaches help to retard population growth, they cannot halt growth by themselves (as reality has painfully demonstrated). The two-credit plan, in contrast, would halt population growth within a year, and then rapidly reverse it. It is a genuine cure, not a palliative. And it would permanently prevent future population growth.

Even in the absence of a two-credit system, any community could still adopt effective antigrowth policies. For example, any village, town, or city could make the goal of population stabilization part of its master plan. It could then enforce that policy by refusing to issue building permits for new housing or commercial development. Let their motto be "if it isn't built, they can't come."

> Why not consider the possibility that a city, like a man or woman or tree or any other healthy living thing, should grow until it reaches maturity—and then stop?
>
> —Edward Abbey[5]

Local governments could also ensure that their school systems educate students about demography, ecology, exponential growth, sex, and contraception. And they could make long-term contraception available at no cost to teens, the unemployed, and the poor. The community could also reform its tax system, so that residents who have more than two children would pay higher taxes to offset the additional educational costs that these extra children impose upon the community. By taking these simple measures, an enlightened community could halt, and then reverse, population growth within its borders. Other communities would observe the benefits and follow suit. Before long, pressure would mount to adopt a nationwide system to reduce population size. At that point, the two-credit system would be the logical choice. But none of this can happen until people are informed about the harmful effects of population growth and demand change.[6]

> He who publishes a moral code, or system of duties, however firmly he may be convinced of the strong obligation on each individual strictly to conform to it, has never the folly to imagine that it will be universally or even generally practised. But this is no valid objection against the publication of the code. If it were, the same objection would always have applied; we should be totally without general rules; and to the vices of mankind arising from temptation would be added a much longer list, than we have at present, of vices from ignorance.
>
> —Thomas Malthus[7]

Notes

1. Interesting firsthand accounts of The Farm can be found in the book *Voices from The Farm: Adventures in Community Living*, ed. Rupert Fike (The Book Publishing Company, 1998).

2. See http://www.pewsocialtrends.org/2015/05/07/childlessness-falls-family-size-grows-among-highly-educated-women/. As of 2014, by the end of their childbearing years, 15 percent of American women had no children, 18 percent had one, 35 percent had two, 20 percent had three, and 12 percent had four or more. There is a nice graph in this article that compares these percentages to 1976. The percentage of no-child and one-child families has gone up since then, and the percentage of 3-child and 4-child-or-more families has gone down.

3. Ibid.

4. For information about the percentage of US environmentalists see http:// www.pewsocialtrends.org/2014/03/07/millennials-in-adulthood/#social-and- religious-views.

5. Edward Abbey, *One Life at a Time Please* (New York: Henry Holt & Co.), 60.

6. People with computer programming skills who are interested in further exploring the two-credit system could incorporate it into software programs that investigate scenarios of the future, such as the World3 model used in *Limits to Growth*. This would further demonstrate the practicality and effectiveness of the two-credit system.

7. Donald Winch, ed., *An Essay on the Principle of Population*, (Cambridge University Press, 1992), 225.

Conclusion

It is really time now to try something else.
—Thomas Malthus[1]

Unrestricted breeding produces hunger, violence, and disease. It destroys the beauty and abundance of the world and denies a good life to future generations. All who love life and care about posterity need to resist it. The credit-based population-stabilization plan presented in this book is a straightforward way to help create a smart, ethical, and compassionate society—one that would have a long future. It is a plan that can swiftly and painlessly reverse population growth and accomplish this without imposing a ceiling on the number of children a couple can produce. Moreover, it would not oblige anyone to use contraceptives or have abortions. At every turn, the plan allows people to make choices.

Yet some will reject this plan out of hand, either because they have persuaded themselves that overpopulation is not a problem or because they want to produce more than two children at society's expense. Some may also object because they think there are better ways to reduce population size. If so, they should share their ideas with us so we can compare them to the two-credit plan with respect to simplicity, practicality, comprehensiveness, compassion, freedom, and fairness.

If you have better, bring it out; if not, give in.

—Horace[2]

Even when a population begins shrinking, a population-stabilization program would still be worthwhile. For one thing, it would bring population size under rational control and prevent the population from ever growing too large (or too small) in the future. It would also legally recognize everyone's reproductive rights, allowing those rights to be sold or donated. The plan would also prevent selfish individuals from treating their community and its resources as an unregulated commons that they could exploit without regard for others. Finally, the plan offers a rational and fair way to handle immigration, resolving once and for all the problem of resident illegal immigrants.

> Whatever lies ahead, we know its main dimensions will emerge in the next two decades. The global economy is already so far above sustainable levels that there is very little time left for the fantasy of an infinite globe. We know that adjustment will be a huge task. It will entail a revolution as profound as the agricultural and industrial revolutions. We appreciate the difficulty of finding solutions to problems such as poverty and employment, for which growth has been, so far, the world's only widely accepted hope. But we also know that reliance on growth involves a false hope, because such growth cannot be sustained. Blind pursuit of physical growth in a finite world

ultimately makes most problems worse; better solutions to our real problems are possible.

—Donella and Dennis Meadows,
and Jørgen Randers[3]

Notes

1. Thomas Robert Malthus, *An Essay on the Principle of Population* (1798), book 4, chapter 3.

2. This line from Horace was quoted by Montaigne in *Michel de Montaigne: The Complete Works*, trans. Donald M. Frame (New York: Alfred A. Knopf, 1943), 397.

3. Donella Meadows et al., *Limits to Growth: The 30-Year Update* (White River Junction, VT: Chelsea Green Publishing, 2004), 12.

Some Informative Websites

albartlett.org. This is an excellent website for gaining an understanding of exponential growth. The late Dr. Bartlett's videos are educational and entertaining, and his voice is a pleasure to listen to.

americanimmigrationcouncil.org. This is the website of the American Immigration Council. It clearly explains the US immigration system.

biologicaldiversity.org. Website of the Center for Biological Diversity, based in Tucson, Arizona. This is one of the few environmental organizations enlightened enough to call for reducing population size, ending animal exploitation, and reducing consumerism: "Human population growth and overconsumption are at the root of our most pressing environmental problems, including the wildlife extinction crisis, habitat loss and climate change."

capsweb.org. Website of Californians for Population Stabilization (CAPS). This organization works to inform Californians about the dangers of population growth. California is America's most populous state, due in large part to mass immigration from Latin America and the higher birth rates of Latinos.

cis.org. Website of the Center for Immigration Studies. Good source of information about US immigration.

cluboforme.org. The Club of Rome is a think tank that played a crucial role in calling attention to overpopulation. It sponsored the influential 1972 book *Limits to Growth*.

davidsuzuki.org. This is the website of the David Suzuki Foundation. The Foundation's goal is to conserve nature and create a sustainable Canada. Although David Suzuki has spoken out forcefully against overpopulation, I find no mention of overpopulation on this website.

donellameadows.org. This is website of the Donella Meadows Institute. The late Donella Meadows along with her husband, Dennis Meadows, and Norwegian scholar Jørgen Randers are the authors of the important and influential book *Limits to Growth*.

earthisland.org. This is the website of the Earth Island Institute (founded by David Brower), which publishes the *Earth Island Journal*. This journal examines environmental problems comprehensively and honestly.

greenvegans.org. This website was founded by Will Anderson, author of *There is Hope: Green Vegans and the New Human Ecology*. This book emphasizes the importance of reducing consumerism and halting both population growth and animal exploitation if we are to have a decent future.

growthbusters.org. Growthbusters is an outstanding podcast featuring interviews conducted by Dave and Stephanie Gardner with experts on overpopulation, climate change, and overconsumption. Check out Dave's DVD: *Growthbusters*.

grist.org. Good source of information on climate change and other environmental problems. However, these problems cannot be solved without reducing population size, a fact ignored by this website.

guttmacher.org. Website of the Guttmacher Institute. This research organization is one of the best sources of information on reproductive health worldwide.

ippf.org (and fosfeminista.org). International Planned Parenthood Federation is a charity that provides reproductive services around the world. It is composed of 141 member associations in 152 countries, with another 24 partners working in 19 countries. It was founded in 1952 in Bombay by Margaret Sanger and Lady Rama Rau at the Third International Conference on Planned Parenthood. In 2021, the Western Hemisphere branch joined with IWHC and CHANGE to form the umbrella organization Fòs Feminista. (*Fòs* being Haitian Creole for force.)

mahb.stanford.edu. Website of the Millennium Alliance for Humanity and the Biosphere. It aims to "connect activists, scientists, humanists, and civil society to foster global change."

msichoices.org. Marie Stopes International is an organization founded in England that provides contraception and abortion services worldwide.

npg.org. Website of Negative Population Growth. NPG works to lower the US fertility rate and reduce US immigration.

numbersusa.com. This is the website of NumbersUSA Education & Research Foundation, which focuses on one subject: US immigration. The Foundation seeks to educate opinion leaders, policymakers, and the public of the need to reduce immigration to a level commensurate with a "high degree of individual liberty, mobility, environmental quality, worker fairness and fiscal responsibility." The founder and president is Roy Beck, a veteran journalist, author of four national policy books, and a national lecturer. His most recent book, *Back of the Hiring Line*, is excellent .

populationmedia.org. This is the website of Bill Ryerson, who has successfully led efforts to use TV and radio serial dramas in poor countries to encourage small family size.

populationinstitute.org. A major organization working to provide the world's people with birth control services.

prb.org. The Population Reference Bureau (founded in 1929) is an excellent resource for population information about many countries.

steadystate.org. This is the website of the Center for the Advancement of the Steady State Economy (CASSE), founded by Brian Czech.

thecarbonunderground.org. Excellent source of information on regenerative agriculture and the critical role it will play in restoring atmospheric carbon to the soil.

transitionnetwork.org. Website of the Transition Network.

usinc.org. This is one of the organizations at the forefront of opposition to mass immigration to the United States. This site provides accurate immigration information.

worldpopulationbalance.org. The website of World Population Balance, founded by David Paxson. "World Population Balance has carried the population stabilization message to hundreds of thousands of people in ten states through hundreds of live presentations and media interviews. These include colleges, universities and high schools, church organizations, service clubs and other special conferences and groups." This is an outstanding organization.

Annotated Bibliography

Among the many excellent books dealing with overpopulation, environmental protection, and public health, the following stand out.

Abbey, Edward. *Desert Solitaire*. New York: Ballantine Books, 1968. One of Abbey's best-loved books. The University of Arizona Press has a nice hardcover edition with a 1987 preface by Abbey.

——. *The Monkey Wrench Gang*. Great writing, great gang, great fun, great goal. Dream Garden Press has the best edition (illustrated by the artist R. Crumb). 1975, 2008.

——.*One Life at a Time, Please*. New York: Henry Holt and Company, 1978, 1983, 1984, 1985, 1986, 1988. Essays by a master of the form.

——. *Abbey's Road*. New York: Penguin Books, 1979. More essays. "His essays, and his novels too, are antidotes to despair." —Wendell Berry

——. *Down the River*. New York: Plume, 1991. Nineteen essays including "Down the River with Henry Thoreau."

——. *The Fool's Progress: An Honest Novel*. New York: Henry Holt and Company, 1988. Abbey's "fat masterpiece." He described it as "a picaresque comedy about life and death, work and play, love and marriage—with a happy ending" (*Postcards from Ed*, p. 257).

——. *A Voice Crying in the Wilderness (Vox Clamantis in Deserto): Notes from a Secret Journal.* New York: St. Martin's Press, 1989. Illustrations by Andrew Rush. Epigrams from Abbey's journal.

——. *Hayduke Lives!* New York, Boston: Little, Brown and Company, 1990. Abbey's last novel, the sequel to *The Monkey Wrench Gang.* The Gang of Four plus Lone Ranger and Earth First!ers defeat Goliath—Goliath being a colossal dragline excavator that walks. I originally thought this machine was a figment of Abbey's imagination, but I later learned that two of these walking behemoths were constructed by Bucyrus-Erie in the late 1960s. Both have since been scrapped. Videos of them walking can be found online.

——. *Confessions of a Barbarian: Selections from the Journals of Edward Abbey,* edited by David Petersen. Boulder, CO: Johnson Books, 1994. Abbey's private side.

——. *Postcards from Ed: Dispatches and Salvos from an American Iconoclast.* Minneapolis, MN: Milkweed Editions, 2006. Abbey's letters, funny and acerbic.

Abbey described his life's work this way: "Why write? How justify this mad itch for scribbling? Speaking for myself, I write to entertain my friends and to exasperate our enemies. I write to record the truth of our time as best I can see it. To investigate the comedy and the tragedy of human relationships. To oppose, resist and sabotage the contemporary drift toward a global technocratic police state whatever its ideological coloration. I write to oppose injustice, to defy power, and to speak for the voiceless" ("A Writer's Credo" in *One Life At A Time, Please.* Henry Holt and Company, 177–78).

Akihiko, Matsutani. *Shrinking Population Economics: Lessons from Japan.* International House of Japan, 2006. The author, a professor at Japan's National Graduate Institute for Policy Studies and a former official of the Ministry of Finance, examines the implications Japan's current population shrinkage. He argues that it has many positive aspects.

Anderson, Will. *This is Hope: Green Vegans and the New Human Ecology*. Winchester, UK and Washington, USA: Earth Books, 2015. The author documents mankind's rapidly escalating assault on the animate and inanimate world. He points out that the two most effective solutions for countering this violence would be to greatly reduce population size and to cease eating animals. The author backs up his facts with a reference section that is 77 pages long! Will Anderson envisions the world's human populations shrinking gradually due to attrition, and he advocates hastening this process by granting a tax deduction of 150 percent for the first child, but no deductions for subsequent children (p. 151). Unfortunately, attrition and tax deductions will not reduce populations fast enough to avoid disaster. What would work—and quickly—is a credit-based system that uses a 1 percent annual reduction rate.

Angell, Marcia. *The Truth about the Drug Companies: How They Deceive Us and What to Do About It*. New York: Random House, 2004. A clear and authoritative explanation of why Americans pay extraordinarily high prices for prescription drugs. The author is a medical doctor and former editor in chief of the *New England Journal of Medicine*.

Atwood, Margaret. *The Year of the Flood*. New York: Nan A. Talese/ Doubleday, 2009. Fictional account of the final years of a collapsing civilization (ours). Very plausible because it is based on what is actually happening. God's Gardeners are mankind's best hope. *Oryx and Crake* (2003) and *MaddAddam* (2013) complete the trilogy.

Beck, Roy. *Back of the Hiring Line: A 200-Year History of Immigrant Surges, Employer Bias, and Depression of Black Wealth*. Arlington, VA: NumbersUSA, 2021. A revealing history of the devastating impact that mass immigration has had on the employment of Blacks in the US and thus their ability to escape poverty.

Bartlett, Albert, with Robert G. Fuller, Vicki L. Plano Clark, and John A Rogers. *The Essential Exponential! For the Future of Our Planet*. Lincoln, NE: University of Nebraska, Center for Science, Mathematics and Computer Education, 2004. The late Albert Bartlett was professor emeritus in nuclear physics at the University of Colorado Boulder. He famously remarked that "the greatest shortcoming of the human race is our inability to understand the exponential function." He devoted his later years to explaining exponential growth to the public. His excellent lectures on this topic can be viewed on YouTube.

Berry, Wendell. *What Are People For?* San Francisco: North Point Press, 1990. One of America's clearest thinkers and finest writers assays our environmental and social problems.

———. *What Matters? Economics for a Renewed Commonwealth*. Berkeley: Counterpoint, 2010. Another fine book.

Boulding, Kenneth. *The Meaning of the 20th Century: The Great Transition*. New York: Harper Colophon, 1964. This is the book in which a system of exchangeable credits to regulate population size was first proposed. Boulding was a brilliant economist, social philosopher, and devout Quaker. See Keyfitz entry below.

Cafaro, Philip. *How Many Is Too Many? The Progressive Argument for Reducing Immigration into the United States*. Chicago: University of Chicago Press, 2015. Cafaro presents a balanced and factual analysis of America's overpopulation problem. He emphasizes the need for the US to end mass immigration because that would be the simplest and quickest way to slow or even halt US population growth. Another fine book by Cafaro is *Thoreau's Living Ethics*.

Campbell, Colin T., and Thomas M. Campbell. *The China Study*. Dallas, TX: BenBella Books, 2006. Written by an eminent scientist and his son, this book is for everyone who wants to learn about healthful diet. It shows how nutrition can prevent and often reverse many chronic diseases. It presents the background science clearly and concisely and includes the author's first-hand account of how powerful economic interests work to suppress the inconvenient truths of nutritional science.

Carroll, Laura. *The Baby Matrix: Why Freeing Our Minds From Outmoded Thinking about Parenthood & Reproduction Will Create a Better World*. Live True Books, 2012. A thoughtful analysis and criticism of pronatalism.

Catton, William R., Jr. *Bottleneck: Humanity's Impending Impasse*. Xlibris Corporation, 2009. The late author, a sociologist, offers valuable insights into our contemporary social and environmental predicaments. He makes the case that too many people consuming too many resources and producing too much pollution will inevitably produce large demographic and economic contractions in the 21st century. This is the "bottleneck" of the title.

Cervantes, Miguel de. *El ingenioso hidalgo don Quijote de la Mancha*. 2 vols. Madrid, 1605–15. All the best of humanity is in this classic tale.

Cohen, Joel E. *How Many People Can the Earth Support?* New York: W. W. Norton & Company, 1995. The author, a mathematician, concludes that the maximum sustainable human population cannot be determined. There is nothing novel about his conclusion: Malthus arrived at it 200 years earlier: "The power of the earth to produce subsistence is certainly not unlimited, but it is strictly speaking indefinite; that is, its limits are not defined, and the time will probably never arrive when we shall be able to say, that no farther labour or ingenuity of man could make further additions to it."

Conly, Sarah. *One Child: Do We Have a Right to More?* New York: Oxford University Press, 2016. The author, a teacher in the philosophy department at Bowdoin College, argues that it is ethical to regulate birth rates in order to reduce the harm of human overpopulation. The logical corollary is that it would be unethical *not* to regulate birth rates.

Daly, Herman E., and John B. Cobb. *For the Common Good: Redirecting the Economy Toward Community, the Environment, and a Sustainable Future.* Boston: Beacon Press, 1994. An informative, wise, and beautifully written book. In importance it ranks with the works of John Stuart Mill and Kenneth Boulding. This book helps explain much that is happening in the world today. Its insights should be absorbed and made part of our outlook. Herman Daly died in October 2022.

Daly, Herman E. *Beyond Growth: The Economics of Sustainable Development.* Boston: Beacon Press, 1996. A highly informative and very well-written book by an outstanding economist. It discusses the problems of our growth-based economy and describes the rational alternative: a steady state economy. Such an economy would be sustainable because it would be scaled to long-term environmental capacities.

Darwin, Charles. Everything Darwin wrote is worth reading—and rereading. These four classics are especially good:

——. *The Voyage of the Beagle.* London, 1839.

——. *On the Origin of Species.* London, 1859

——. *The Descent of Man, and Selection in Relation to Sex.* London, 1871.

——. *The Expression of the Emotions in Man and Animals.* London, 1872.

Diamond, Jared. *Collapse: How Societies Choose to Fail or Succeed*. New York: Viking, 2005. This book presents examples of societies that failed and others that survived (and why). Two of Diamond's other books—*Guns, Germs, and Steel* and *The Third Chimpanzee*—are also first rate.

Dietz, Rob, and Dan O'Neill. *Enough is Enough: Building a Sustainable Economy in a World of Finite Resources*. San Francisco: Berrett-Koehler Publishers, Inc., 2013. This book envisions how a steady state economy would work.

Esselstyn, Caldwell B., Jr., MD. *Prevent and Reverse Heart Disease: The Revolutionary, Scientifically Proven, Nutrition-Based Cure*. New York: The Penguin Group, 2007. An excellent resource for learning how to prevent heart attacks and strokes. Dr. Esselstyn was president of the staff at the Cleveland Clinic.

Everett, Daniel L. *Don't Sleep There Are Snakes: Life and Language in the Amazonian Jungle*. New York: Vintage Departures, 2009. A fascinating account of a "primitive" people who live happily and in harmony with their environment.

Foreman, Dave, with Laura Carroll. *Man Swarm: How Overpopulation is Killing the Wild World*. Live True Books, 2014. http://lauracarroll.com. Another good book on overpopulation by an outstanding conservationist. Sadly, Dave Foreman died in 2022.

Gilding, Paul. *The Great Disruption: Why the Climate Crisis Will Bring on the End of Shopping and the Birth of a New World*. Bloomsbury Press, New York, 2011. The author, who is a former head of Greenpeace, believes it is already too late to halt population growth. He therefore focuses on the need to reduce consumption and increase efficiency in resource use. Although he means well, the author's belief that mankind can be saved by reducing per capita consumption (the multiplicand) while ignoring the need to reduce population size (the multiplier), is naive and wrong. He also fails to see that next to population reduction, the most effective way to reduce harmful environmental impacts would be to adopt a plant-based diet.

Greger, Michael, MD. *How Not to Die.* New York: Flatiron Books, 2015. An excellent up-to-date summary of the scientific evidence on the relationship between diet and health. The verdict of nutritional science is unequivocal: a plant-based diet is best for health and best for the environment.

Hardin, Garrett. *Naked Emperors: Essays of a Taboo-Stalker.* Los Altos, CA: William Kaufmann, Inc., 1982. An excellent collection of smart, thought-provoking essays.

——. *Filters against Folly: How to Survive Despite Economists, Ecologists, and the Merely Eloquent.* New York: Penguin Books, 1985. This book examines the need for literacy, numeracy, and ecolacy and calls attention to the pervasiveness of the "double *C*–double *P* game" (commonizing costs while privatizing profits). Hardin's books are always worth rereading.

——. *Living within Limits: Ecology, Economics, and Population Taboos.* Oxford, UK: Oxford University Press, 1993. Another informative book on overpopulation and related subjects.

——. *The Ostrich Factor: Our Population Myopia.* Oxford, UK: Oxford University Press, 1998. Hardin's final book.

Hartman, Edward C. *The Population Fix: Breaking America's Addiction to Population Growth.* Moraga, CA: Think Population Press, 2006. A well-written and useful book on the overpopulation problem in the United States. It includes political steps that can be taken to counter it.

Heard, Gerald. *Pain, Sex, and Time: A New Outlook on Evolution and the Future of Man.* Rhinebeck, NY: Monkfish Book Publishing Company, 1939. Heard was a brilliant spiritual philosopher whose work and life were highly influential. Some of his recorded talks are available online.

Heinberg, Richard. *The End of Growth: Adapting to Our New Economic Reality*. Gabriola Island, British Columbia: New Society Press, 2011. The author makes the case that we must end our senseless pursuit of ever more consumption and instead direct our efforts toward improving the quality of life.

Heinberg, Richard and David Fridley. *Our Renewable Future: Laying the Path for One Hundred Percent Clean Energy.* Washington, DC: Island Press, 2016. This book explains what we can and should do to "decarbonize" our economy.

Hofmann, Albert. *LSD: My Problem Child; Reflections on Sacred Drugs, Mysticism, and Science*. Santa Cruz, CA: Multidisciplinary Association for Psychedelic Studies, 2009. The wise father of LSD and psilocybin tells the story of his discoveries and how they transformed his life and the world. At the age of 100, Hofmann wrote "I hope that LSD provides to the individual a new worldview, which is in harmony with nature and its laws."

Holland, Julie, MD. *Good Chemistry: The Science of Connection from Soul to Psychedelics.* New York: HarperCollins, 2020. The author, a psychiatrist, presents the latest scientific research into the physical and spiritual healing powers of cannabinoids and psychedelics.

Hudson, Michael. *Killing the Host: How Financial Parasites and Debt Destroy the Global Economy.* ISLET-Verlag, 2015. www.michaelhudson.com and www.islet-verlag.de. This is a valuable book for understanding modern economies and the parasitic role of the financial sector.

Huxley, Aldous. *Island*. New York: HarperCollins, 1962. *Island* was Huxley's final work. He used this fictional tale to present his ideas on what a better society would be like and to warn of the threats such a society would face. This book was Huxley's answer to his earlier (1932) *Brave New World*. According to his wife, Laura, "Aldous was appalled I think (and certainly I am), at the fact that what he wrote in *Island* was not taken seriously. It was treated as a work of science fiction, when it was not fiction, because each one of the ways of living he described in *Island* was not a product of his fantasy, but something that had been tried in one place or another, some of them in our own everyday life" (*Moksha: Aldous Huxley's Classic Writings on Psychedelics and the Visionary Experience*, p. 266).

Keyfitz, Nathan. "Kenneth Ewart Boulding: 1910–1993." Washington, D National Academies Press, 1996. A biographical memoir of an outstanding economist and social philosopher. Can be read online at http://www.nap.edu/html/biomems/kboulding.pdf.

Kohr, Leopold. *The Breakdown of Nations*. Devon, England: Green Books Ltd., 2001 (originally published in 1957 by Routledge and Kegan Paul). This is a little known but important work. Kohr was an economist who believed that the key to achieving enduring international peace was for big states to break up into small states so that all states would have roughly equal population and power. Naturally, he doubted that this would ever take place, because big states are loath to relinquish their power. Kohr seems not to have imagined that populations could *deliberately* become small enough that conflict over resources would disappear.

Lao Tzu. *Tao Te Ching: A New Translation*, translated by Sam Hamill. Boston: Shambhala Publications, Inc., 2007. Edward Abbey praised the *Tao Te Ching* as "a tiny little open-ended giant of a book" (*Hayduke Lives!* page 167). Sam Hamill is an accomplished poet and scholar of ancient Chinese.

Lao Tzu. *The Tao Te Ching of Lao Tzu*, translated by Brian Browne Walker. New York: St. Martin's Press, 1995. I don't know Chinese, but after comparing this translation with a few others, I sense that it is one of the very best. The book is also beautifully designed.

LeBlanc, Steven A., with Katherine E. Register. *Constant Battles: Why We Fight*. New York: St. Martin's Press, 2004. An outstanding book by an archaeologist making the case that overpopulation has been the driver of war throughout mankind's existence.

Lucretius. *Lucretius: The Way Things Are: The De Rerum Natura of Titus Lucretius Carus*, translated by Rolfe Humphries. Bloomington, IN: Indiana University Press, 1968. In this long poem, Lucretius expounds the philosophy of his hero, the Greek philosopher Epicurus. It is remarkable how many of the ideas of Epicurus agree with those of modern science, including the atomic structure of matter and the conservation of mass. Epicurus rejected supernatural explanations for "the way things are."

Malthus, Thomas Robert. *An Essay on the Principle of Population*, 1798 (and later editions). This insightful and highly influential work was the first to address the problem of overpopulation. A must read for anyone interested in population issues.

Meadows, Donella H., Dennis Meadows and Jørgen Randers. *Limits to Growth: The 30-Year Update*. White River Junction, VT: Chelsea Green Publishing Company, 2004. An indispensable guide for understanding the harm mankind is doing to the earth and what can be done to avert disaster. The third edition (like the first edition published in 1972) discusses several different scenarios of the future based on sophisticated computer models.

Mill, John Stuart. *Principles of Political Economy*. New York: Prometheus Books, 2004. Originally published in 1848 (revised in the 1860s). This book was the most highly regarded 19th-century treatise on economics. It elucidates many economic processes. Mill was a genius, and it is a great pleasure to read him.

——. *On Liberty*, 1858. This superb essay can be read online at http:// www. bartleby.com/130/1.html.

——. *The Subjection of Women*, 1869. Early and cogent argument for equal rights for women. Can be read online at https://www.gutenberg.org/ files/27083/27083-h/27083-h.htm.

Montaigne, Michel de. *Michel de Montaigne: The Complete Works*, translated by Donald M. Frame. New York: Alfred A. Knopf, 1958. Montaigne was a wise and entertaining thinker, who was deeply versed in the Greek and Roman classics.

Montgomery, David R. *Dirt: The Erosion of Civilizations*. Berkeley: University of California Press, 2007. A valuable book showing the critical role that soil nutrient exhaustion and erosion have played in human history.

Morland, Paul. *The Human Tide: How Population Shaped the Modern World*. New York: Hachette Book Group, 2019. Morland, a demographer, focuses on demographic transitions. Although this book is informative and interesting, it contains a massive hole: The author never once mentions overpopulation, sustainability, climate change, or mass extinction. In Morland's world, overpopulation and its consequences can be ignored, because fertility rates are falling in many countries. He thinks these falling rates imply that mankind has escaped the "Malthusian trap." What Morland doesn't understand is that the trap has already sprung. It is now inevitable that many populations will suffer mass premature death until their numbers decline to carrying capacity (which will be less than it was before due to environmental degradation). Morland also fails to see that the only sensible response to the Malthusian trap is a rational program of population control.

Morrison, Reg. *The Spirit in the Gene: Humanity's Proud Illusion and the Laws of Nature*. Ithaca, NY: Cornell University Press, 1999. Essential reading for anyone trying to understand human nature and mankind's future. Morrison argues that our reproductive behavior is controlled by our genes through our instincts. And he believes that reason cannot prevail over instinct. Having abandoned hope that humanity can voluntarily reduce its numbers, he believes the earth's own self-regulatory capacity will finally step in to reduce human numbers.

Orlov, Dmitry. *The Five Stages of Collapse: Survivor's Toolkit*. Gabriola Island, British Columbia: New Society Publishers, 2013. The author's thesis is that societal collapse occurs in five stages, starting with financial collapse (which can happen almost overnight), followed by commercial collapse, governmental collapse, social collapse (the breakdown of basic social services), and finally cultural collapse (the breakdown of civilized norms, resulting in chaos and violence). The author's personal experience with financial, commercial, and political shocks in Russia in the early 1990s has informed his understanding of the first three processes.

Ornish, Dean. *The Spectrum: A Scientifically Proven Program to Feel Better, Live Longer, Lose Weight, Gain Health*. New York: Ballantine Books, 2007. Everything the title promises.

Piketty, Thomas. *Capital in the 21st Century*, Cambridge, MA: Belknap Press, 2014. Piketty, an economist, elucidates the growing inequality in modern societies, and the threat it poses.

Potts, Malcolm, and Roger Short. *Ever since Adam and Eve: The Evolution of Human Sexuality.* United Kingdom: Cambridge University Press, 1999. A highly informative book that covers the history and science of human sexuality. Malcolm Potts is a human reproductive scientist and professor at the School of Public Health of the University of California, Berkeley. He is the founding director of the Bixby Center for Population, Health, and Sustainability. His coauthor, Roger Short, is the Wexler Professorial Fellow in the Royal Women's Hospital at the University of Melbourne, Australia. Short began his career as a veterinarian and went on to study many aspects of human and animal reproduction.

Potts, Malcolm, and Thomas Hayden. *Sex and War: How Biology Explains War and Terrorism and Offers a Path to a Safer World.* Dallas, TX: BenBella Books, 2008. A superb work that elucidates the link between overpopulation and group violence, tracing the male attraction to team aggression to our chimpanzee-like ancestors. For those wishing to understand human violence, this book is indispensable.

Raine, Adrian. *The Anatomy of Violence: The Biological Roots of Crime.* New York: Vintage Books (a division of Random House), 2013. The author, a psychologist, presents the latest scientific findings on how biology and culture interact to produce violent, antisocial people.

Rieder, Travis N. *Toward a Small Family Ethi How Overpopulation and Climate Change Are Affecting the Morality of Procreation.* Springer Nature, 2016. Rieder uses philosophy, his specialty, to examine the ethics of reproduction in an overpopulated world. Not surprisingly, he concludes that people should have few or no children. Although he avoids the controversial issue of immigration, his arguments in favor of less reproduction would apply equally to immigration.

Sapolsky, Robert M. *Behave: The Biology of Humans at Their Best and Worst.* New York: Penguin Books, 2017. An informative and thought-provoking book on how human behavior is governed by the interaction of biology and culture. The author is a neuroscientist and primatologist.

Shragg, Karen I. *Move Upstream: A Call to Solve Overpopulation.* Minneapolis: Freethought House, 2015. The author's aim is to persuade activists who struggle against the fallout from overpopulation (the "downstream issues") to finally start working against overpopulation itself (the "upstream issue"). Most of the book is an exploration of the great silence that surrounds the topic of overpopulation. Oddly, the author herself is silent about the main driver of US population growth: mass immigration. No matter. This is an excellent book that deserves a wide readership. The author invites us to "imagine a world where all of the caring celebrities, the human rights organizations, the conservation groups, the peace groups and the artists all took on overpopulation as their overarching message. I believe that the solutions would come pouring in and become quite transparent, especially if journalists got on board."

Smith, Dick. *Dick Smith's Population Crisis: The Dangers of Unsustainable Growth for Australia.* Melbourne: Allen & Unwin, 2011. An excellent book that focuses on the overpopulation problem in Australia.

Soddy, Frederick. *Wealth, Virtual Wealth, and Debt: The Solution to the Economic Paradox.* London: George Allen & Unwin, Ltd., 1926. After winning the Nobel Prize for chemistry in 1921, the author turned his brilliant mind to economics. This informative book is the result.

Stern, Andy, with Lee Kravitz. *Raising the Floor: How a Universal Basic Income Can Renew Our Economy and Rebuild the American Dream.* New York: Public Affairs, 2016. This bellwether book explores the impending crisis of mass unemployment and recommends a universal basic income as the practical and morally right solution.

Thoreau, Henry David. "Resistance to Civil Government" (posthumously renamed "On the Duty of Civil Disobedience"), 1849. Influential essay on our responsibility not to contribute to government-sponsored evil.

——. *Walden,* 1854. A literary masterpiece and guide to living. It is the original (and best) introduction to deep ecology.

———. *The Journal*, 1837–1861. A 14-volume literary and philosophical treasure. An abridged version edited by Damion Searls was published by the New York Review of Books in 2009.

Weisman, Alan. *Countdown: Our Last, Best Hope for a Future on Earth?* New York: Back Bay Books/Little, Brown and Company, 2013. A highly informative examination of overpopulation. The author describes countries that have been successful at reducing overpopulation and others that are failing.

Yang, Andrew. *The War on Normal People: The Truth about America's Disappearing Jobs and Why Universal Basic Income Is Our Future.* New York: Hachette Books, 1999. The author, a candidate for US president in 2020 and for mayor of New York in 2021, clearly and persuasively presents the case for a universal basic income (UBI). What Yang does not acknowledge is that such a program would need to be accompanied by a system of population control. Otherwise, additional income will result in harmful population growth.

NOVEMBER SURF

Some lucky day each November great waves awake and are drawn
Like smoking mountains bright from the west
And come and cover the cliff with white violent cleanness: then
    suddenly
The old granite forgets half a year's filth:
The orange-peel, egg-shells, papers, pieces of clothing, the clots
Of dung in corners of the rock, and used
Sheaths that make light love safe in the evenings: all the droppings
    of the summer
Idlers washed off in a winter ecstasy:
I think this cumbered continent envies its cliff then…But all
    seasons
The earth, in her childlike prophetic sleep,
Keeps dreaming of the bath of a storm that prepares up the long
    coast
Of the future to scour more than her sea-lines:
The cities gone down, the people fewer and the hawks more
    numerous,
The rivers mouth to source pure; when the two-footed
Mammal, being someways one of the nobler animals, regains
The dignity of room, the value of rareness.

—Robinson Jeffers

Shams and delusions are esteemed for soundest truths, while reality is fabulous. If men would steadily observe realities only, and not allow themselves to be deluded, life, to compare it with such things as we know, would be like a fairy tale and The Arabian Nights' Entertainments.

—Henry Thoreau, *Walden*

Science, with its intrinsic honesty and its rejection of the supernatural as an answer to real world events, has proved the only medium in history capable of linking women and men of all cultures and all races in a common understanding of the real world.

—Malcolm Potts and Thomas Hayden, *Sex and War*

Science is international; or supra-national, if you wish. This is its glory and its strength.

—Garrett Hardin, *Filters Against Folly*

Folks, it's time to evolve.

—Bill Hicks

About the Author

The author was born in 1949 and grew up in Detroit. As a young man, he declined to be conscripted into the army and spent four years as a fugitive. Later, he became an aircraft mechanic and then a botanist. He has written one other book: *Psilocybin Mushroom Legal Defenses*. His Polish surname comes from *brod* (ford) or *broda* (beard) and is pronounced with the accent on the middle syllable (long *o*).

www.ingramcontent.com/pod-product-compliance
Lightning Source LLC
Chambersburg PA
CBHW041254040426
42334CB00028BA/3010